Advance Praise For THE REP.

"Ancient catastrophes triggered a power outage that left students of history groping in the dark. Neville Raymond turns the lights back on. His bulldozer of a book moves back and forth between cosmology and psychology to clear away the mudslides that block our pathways to love."

Toni Toney, *author of **EcoDiet**, and creator and founder of the Ecotarian Revolution*

"Neville Raymond brilliantly expands the horizon of childhood trauma to encompass the collective trauma of the human race. The future health and well-being of humanity revolves around the principles found in this book. Passionate, eloquent and insightful beyond belief, **The Reparenting Revolution** provides trail-blazing approaches we can adopt to remake the world in the process of remaking ourselves."

Dr Sarah Larsen, *MD and Medical Intuitive, Public Speaker, Host of Miracle Makers podcast, Radio/TV, Consultant to High Impact Entrepreneurs and Visionaries*

"This book starts an astonishing new chapter on the human race. Read it and let's start a real conversation about the next step in enlightened parenting. Our kids and the future of our world demand it."

Richard Greene, *Civics Educator, Public Speaking Coach for Presidents and CEOs, Author of **Words That Shook the World***

THE
REPARENTING
REVOLUTION

From a Power-Based

To a Love-Based Society

NEVILLE RAYMOND

Los Angeles CA

First Edition 2018

Raymond, Neville, 1949, September 8 -

The Reparenting Revolution

Book Cover by Carl Graves, Extended Imagery

Website: catwebsites.com

Published in the USA

IngramSpark (ingramspark.com) & Amazon CreateSpace

1st Edition/ October 2018

SBN - 978-0-692-19400-3 Paperback

nevilleraymond.com

DEDICATION

To my wife and two sons –

may they live long enough to rejoice in the family of man.

CONTENTS

PREFACE

Men die because they cannot link together the beginning to the end.

— **Alcmaeon of Croton (c. 500 B.C.)**

The Question

Wouldn't it be something if the mass shootings that are a fixture of modern-day life in America were a holdover from a time when the earth was a shooting gallery for the heavens?

Or if unmanned aerial vehicles make a killing fields of the earth today because once upon a time man revered as gods the celestial forces that rained death and destruction from on high?

Wouldn't it boggle the mind to know that the catastrophic fallout from comets long dreaded as bad news continues to inform the bad news we read about in the morning papers?

Or that the most significant impact of fireballs from the sky are not giant craters in the ground but gaping holes in the ground of our being?

Wouldn't that radically alter our perception of how the world is organized?

Once in a while a book flips a switch in our brains and changes the way we think. *The Reparenting Revolution* is such a book.

And it begins by asking, Why is mankind such a nervous wreck?

Love is the principle by which we organize the prosocial order of our being.

Why then has power been the organizing principle for society for thousands of years?

We do justice to our humanity by loving it unconditionally.

Why then does our system of justice have more in

common with a supernatural dispensation of hell and damnation than a natural one of nurture and affection?

Our need for love, our affirmative zest for life, is rooted in our mammalian biology.

Why then our cultural obsession with death, our compulsive mania for destruction?

A logical place to start is what we know from first-hand experience.

A soul-searing history of battery, rape, abandonment and neglect, is not good for the development of children. We know it breeds social deviance and all kinds of sociopathic behavior.

What of human history as a cavalcade organized by the four horsemen of the apocalypse?

Doesn't it occur to us to wonder what kind of universal background of death and destruction could account for history as such an encyclopedic dossier of crimes against humanity?

If no mass murderer ever had a happy childhood, what can be extrapolated from our collective history of serial wars, massacres, conquests and enslavements on a scale unimaginable?

Can we safely deduce that mankind was terrorized from the cradle?

And if mankind had a childhood from hell, who or what was responsible?

Who were the authority figures who acted like abusive parents? Who were the martial arts teachers of mass murder and mayhem? Who were the demented role models that demanded that grown-ups be immolated as burnt offerings and children be initiated in crucibles of fire?

In whose name did rulers oppress, exploit and enslave

their fellow man? And to the greater glory of what beings did they launch endless cycles of invasions, conquests and wars?

It's time to draw the conclusion that has been building up for centuries.

Our preoccupation with judgmental forms of violence and destruction in the present is a sure-fire sign of an apocalyptic Day of Judgment in the past.

The human condition is seen clearly through the optics of childhood trauma.

There is our trauma as individuals. And then there is our trauma as a species.

And this book is the first attempt to synthesize the two to ignite a revolutionary new understanding of history.

The Answer

The premise of *The Reparenting Revolution* is as stark as it is simple.

Our species suffered a series of near-death experiences.

Ancient myths and legends speak of a Day of Destruction.

Modern findings in geology and astrophysics prove it.

And yet no reckoning has been made of its horrific toll on the human psyche.

The handful who survived these earth-shattering catastrophes were equal to the task of repopulating the earth – but they were in no shape to cope with their nerve-shattering trauma.

And the ramifications of their trauma-crazed legacy have gone unnoticed and unexplored.

Those who survived the cosmic firebombing of earth were *de facto* parents of humanity – mother and father of the never-ending barrage of assaults they passed on to the family of man.

They could not very well remain grounded in their beings under such an insupportable weight of trauma. And so they split off from their humanity to identify with the celestial forces that aggressed against them, idolizing them as objects of worship known as gods.

The flip side of man's faith in gods is a corrosive distrust of human nature. Supernatural commands to please gods displace man's natural instincts to find joy and pleasure in his being.

A reign of terror instituted by doomsday gods launches the first wave of man's inhumanity to man in the form of infanticide, blood sacrifice, initiation rites, cannibalism, and slavery.

The institution of kingship is a spin-off from this primal split in our psychology.

Man is no longer ruled by impulses that spring from his humanity.

He is by ruled by men who, as self-avowed gods, would be ruled certifiably insane today.

They propagated their rule by making over the social order in their schizoid image.

As deranged beings who were split off from themselves, they necessarily presided over a society of estranged beings who were cut off from each other.

This is the ancient origin of the power principle – *divide and conquer.*

The power structure has always perpetuated itself by breeding division and discord, conflict and strife in the social body.

The wrath-crazed judgments of gods are channeled through vengeful judgments of men.

A political order founded on a penal or punitive dispensation gives rise to a second wave of man's inhumanity to man in the form of incarceration, torture, executions, genocide and war.

We pride ourselves on a separation of church and state.

The dark, ugly truth is religion and politics are joined at the hip.

The belief in gods would have died a natural death – were it not enmeshed in a state of co-dependence with a power elite with grandiose delusions of godhead.

Once the psyche is branded with a doomsday imprint,

religion makes a show of saving us from the wrath of gods, but actually controls us through a manufactured fear and terror of gods.

Political authority is cut from the same whole cloth as religion. It too puts on a great show of protecting us from the wrath of our fellow man but in actuality controls us through a manufactured fear and terror of our fellow man.

And so it is that the trail of blood and tears that hark back to religious man as bedfellow of the gods broadens into the terrors of history – the mass deportations and inquisitions, the global privations, purgations and conflagrations that trace back to the politics of men playing god.

The Solution

Man evolved from the apes, but our religious and political order has always been a crash adaptation to the monstrous gods that terrorized man in form of comets, planets and meteors.

Through a cosmic war of the worlds, a loving species designed by nature for cooperation and community becomes an embattled one whose history is defined by competition and carnage.

A parental model for raising the weak and small to be sovereign beings in their own right is upended by a predatory model that is bent on keeping people in peonage and nonage.

The key is to understand how cosmic catastrophes dislodged us from our heart center and led to a centralization of political rule. Only then can we run the process in reverse to demystify the power principle, break the spell of authority, and recover the loving essence of our humanity.

In a practical sense, the book is a suicide helpline for the human race. Its stream of heart-opening insights and groundbreaking observations is poured out in the hopes of talking mankind kind down from the ledge of a dystopian future to live out the utopian possibilities of progress.

For there we are, even after all this time, still trapped by racial memories of frantically jumping off the deep end, as the burning edifice of nature slides apocalyptically into the abyss.

A celestial sword of Damocles may as well be hanging over us, for all the secure foothold we have on earth. Life is a constant struggle to survive for most of us. The freedom to live to the full is beyond our reach. Our potential is squelched and squandered as a matter of course. Our humanity is mutilated and maimed by a battery of ritual ordeals. And our idea of keeping order is to weaponize the judgment of gods and return us to the stardust from which we came.

Support for a power-mad structure still comes from old-school methods of parenting that recycle a traumatic past by conditioning each generation to harden and toughen itself against it.

We don't need violent upheavals to revolutionize the world because they already have!

The breakaway moment will come with a revolution in reparenting – a grassroots movement that gifts every human being on this planet with his or her birthright of unconditional love.

As we release our collective trauma and recover our sanity, the realignment with our humanity triggers at last a seismic shift from a power-centered paradigm to a love-centered one.

The latest developments in anthropology, geology and cosmology confirm the extraterrestrial violence that is the foundation of civilization.

The latest developments in psychology and social science confirm something altogether different – the intrinsic love that is the foundation of humanity.

Now that we know that the dominant values of the civilized order – our worship of gods, our rule by power-mad men masquerading as gods, our cult of authority figures, our ideology of domination, our atavistic confusion of justice with punishment – all derive from a hideous breakdown of cosmic order, we can triage the emergency care our wounded race is dying to receive.

The cosmic order repaired itself a long time ago.

All that remains is to rebuild the integral order of humanity.

A race that moves mountains can move mountains of pent-up trauma out of its way.

A species that mines its cognitive intelligence to triumphantly bend nature to its will can source its emotional intelligence to therapeutically realign society with its heart and soul.

We have always known a culture of empathy and compassion, a society of cooperation and mutual support is what it takes to fulfill our long-awaited destiny – *the brotherhood of man.*

A cosmology of violence from the birth of history made our killer civilization what it is.

A biology of love from the birth of life makes us the joyous species we are.

I. MAYDAY MORALITY

In summary form, it could be said, for primitives, the End of the World has already occurred....Myths of cosmic cataclysms are extremely widespread. They tell how the World was destroyed and mankind annihilated except for a single couple or a few survivors.

— Mircea Eliade, **Myth and Reality** p. 54

Man must have been subjected to some particularly overwhelming experiences to have been led to introduce such cruel practices into his life.

— Adolphe E. Jensen, **The Violence of the Sacred,** René Girard p. 93

THE REPARENTING REVOLUTION

1. WHO ARE YOU OR I TO CHANGE THE WORLD

Tell people that you're writing a book and the first question out of their mouths is: "What's it about."

The answer: "It's about our birthright of unconditional love and how we can change the world by giving it to each other."

At this point most people's response is: "Wow, that sounds awesome."

Without knowing it, much less stopping to think about it, people have just been zapped by three profundities.

Unconditional love is our birthright.

We can give it to each other.

And by giving it to each other we can change the world.

People know what you mean by changing the world. You won't get an argument there. I've never heard anyone say, "The world's fine the way it is. Why fix it if it's not broken?" Except, perhaps, my son when he was living the charmed life of a ten year old. In general people have a sinking awareness of a broken world. Start with your personal life. *My boss is a bully. My partner cheats. My kid uses drugs. I lost my home.* Move on to the community. *Traffic is murder. I got mugged. My priest molested me. My councilman is on the take.* Turn to the nation. *Healthcare legislation hijacked by the industry it is intended to reform. Wars in Afghanistan, Iraq, Syria. Jobs moving abroad. A bankrupt treasury.* Last but not least the world. *World poverty. Terrorism. Eco-disaster.*

Is that all? You must be wondering what we have got ourselves into! This vast, complex world is a handful – even for those into whose hands we have entrusted it. How can we go about changing it without getting in over our heads? Where do we begin? Is it not better left alone? And who am I, or you, or anybody else to dare to change the world?

Who am I? Look in the mirror. I am made in your likeness. My needs are your needs. I want family, community, a better life for my kids. I want freedom to live my passion and follow my bliss. I want to be able to say life is fair because society is just. I want people to live in harmony with the music in their hearts. I want peace on earth.

There are billions of us, wanting the same things out of life, all instantly relatable to each other. We know who we are. The question is not who am I or you to change the world. The question is who are all those who would keep it as it is? Who are those who have their signals crossed, who love the *world* unconditionally as it is, instead of the *people* in it? Who are those enamored of a society designed like the poles of a magnet, its lines of force eternally polarized between those who take all they can get and those taken for everything they have got? Who are those so accepting of a world of conflict and strife that they run about preaching that greed is healthy, competition breeds excellence, punishment is the antidote to crime, poverty is a sign of moral bankruptcy, the wealth pyramid is a tribute to survival of the fittest, and war is a way of settling human differences?

2. TWEEDLEDUMB VS. TWEEDLEDUMBER

Change doesn't occur in a vacuum. There is a word for the atmosphere in which it breathes, the element in which it lives. *Contrast*. It is not the cinematic contrast between good guys and bad guys. It is a real-life contrast between how bad things are – and how good they could be. Could you even know how bad things are in the *present* if you have no way of knowing how good they were in the *past* or how great they could be in the *future*? Without that contrast there is no basis for transitioning from one to the other.

In fact, the more glaring the contrast, the more urgent the need for change. The brighter the future we envision, the dimmer our view of the present. The clearer we see what a wonderful world it could be for every one of us, the weaker our attachment to the views of those who want it to be a wonderful world for them alone. For consumers it is all about chasing after flashy promises of *new* and *improved*. For voters, it is all about settling for the dubious claim that we already have the *best* system available. In fact, if you want to fence people in with the *status quo* in perpetuity, just hypnotize them into believing that the grass is always going to be brown, dry and withered on the other side.

Paradoxically, then, the study in contrasts can work against change just as it can work in favor of it. In fact, when the contrast is framed in less than flattering terms, it can serve as one of the greatest stumblingblocks to progress. If a particular system works well for the benefit of a particular class, why wouldn't it do everything in its power to convince us that this is as good as it gets? If most of its capital is invested in the belief that this is the best of all possible worlds, why shouldn't it promote the idea that any other world would be a step down or a step backward? What is more subversive than for the gilded allure of one political system to be tarnished in comparison to another? That is why our political system often acts like Snow White's stepmother in front of the talking mirror – it cannot tolerate the thought that ours is not the fairest and best system on the planet, and flies into murderous tantrums at the possibility of something better out there.

How's that for Lady Liberty behaving like an autocratic Queen! But then again, Helen Keller saw through her melodramatics long ago. "Our democracy is but a name. We vote? What does that mean? It means that we choose between two bodies of real, though not avowed, *autocrats*. We choose between Tweedledum and Tweedledee."

For voters it is one-size-fits-all. For the powers that be it is one standard for them and one for everyone else. Politicians play fast and loose but one standard they adhere to unfailingly is the double standard. They are gung-ho for wars on dark-skinned peoples, demonized as the forces of darkness. When it comes to the age-old conflict known as class warfare, waged by superrich white people, they run and hide like gutless wonders behind a front of pacifism. They are the Hanoi Janes of the anti-class-war movement, posing for photo-ops on tanks and aircraft carriers while providing aid and comfort to the real enemy – the money power that has done more grievous harm to the American way of life than any Evil Empire.

Politicians are the leading conscientious objectors of class warfare. They send our children to fight the Iraqi or Afghani people, even as they dodge their sworn duty to defend the health, welfare and educational advancement of the American people. They deplore the appeasement policy towards a dictator supposedly bent on world conquest, even as they spinelessly appease the imperialism of money in that relentless march to take over the world that is politely referred to as globalization. They display the vengeful fury of an Old Testament god when it comes to hounding petty criminals who steal from the private till, and display a forgiving spirit worthy of a New Testament god when it comes to corporate scofflaws who loot the public treasury. Nowhere it seems do our public servants practice Christian charity with greater zeal than on behalf of the superrich. If robber barons take the people's coat, politicians are ready to offer them the people's cloak as well. That is how you wind up with a class of have-it-alls and a class of have-nothings.

In the meantime, as the controlled media filters out all meaningful alternatives to a system of plutocracy that is rigged to get bigger and stronger with each passing decade, democracy turns into a game of make-believe where the object is for the players to fight to win by acting as if the only choice on the table is between a *greater* and a *lesser* evil.

Most families who get by – paying the bills, putting aside a little extra for college – know there is room for improvement before life can be as good as it gets. Invite them to get off their chests what ails them and out of their mouths flies a veritable Pandora's box of ills. But the hope at the bottom of their hearts is always the same. It is the hope that there is more to life than the struggle to hold it together. And the only reason they can even bellyache about this world is because in their gut they know a better one is possible.

Yet in a free-market democracy that boasts a numbing array of choices between coffee and toothpaste brands, the daily range of choices for many Americans narrows down to having their houses foreclosed or their cars repossessed. To filling their bellies or their prescriptions. To being out of work with no money to feed and shelter their families or being worked to death with no time to pay attention to their kids, much less nurture the birth of new life. To burning out on a job that pays the bills or being fired up by one that leaves them unsure of where the next paycheck is coming from. To battling terrorists who blow up our infrastructure or fighting bureaucrats who let it fall apart. To voting for a party that lets corporations get away with first degree murder or one that lets them get away with murder in the second. To electing a president who brazenly rapes the American people or one who seduces them with sweet nothings and then betrays them.

Is this the best we can hope to do in the home of the Declaration of Independence and the land of the Bill of Rights? Always having to settle for a lesser of two evils instead of having the freedom and the courage to go after the incontestably greater good?

Instinctively we know better. We cannot keep choosing between shades of evil and expect to show up in our best light. As long as we go on navigating between two more or less equally hazardous wrongs, we are never going to steer the world in the right direction. Life doesn't have to be lived between a rock and a hard spot. It can be lived between an oasis and a safe harbor – a cozy fireplace and a calming spa. We don't have to be caught between the devil and the deep blue sea. We can be torn between sunbathing in a garden of earthly delights and snorkeling in a coral garden under the sea. We may not know exactly how to get from here to there. But in our heart of hearts we know that *here* simply cannot be all there is. And *there* life is everything it can be, and then some.

3. DOMINATING THROUGH EMOTIONAL JUJITSU

As soon as you plump for a world brimming with joy and abundance, the anti-change folks swarm out of the woodwork. The naysayers go by various aliases but the one they like best is *conservative*. Resistance to change is their middle name. They may scandalize the world by cheating on their spouse but they are unswervingly faithful to the status quo. They would do anything to keep things as they are – or jolly them along in a direction where a minuscule fraction of humanity grabs ever larger swaths of our planet earth, while ever larger swaths of humanity are left clinging to a minuscule fraction of it.

Of course, this flies in the face of all commonsense and decency. In a study of primates, researchers found that capuchin monkeys have a built-in sense of equalitarian fairness. If two monkeys get the same reward for a task, all is well. If one monkey is given a less desirable commodity like a cucumber in exchange for a pebble, while another gets a more desirable commodity like a grape without having to hand over a pebble, the body language of the shortchanged monkey is a study in outrage. Hence the age-old dilemma of the ruling class. *They must take a social order that would offend even a monkey's sense of justice and try every which way to pass it off as the natural order of things.*

For a long time their favorite trick was to convince us they are in a superhuman class by themselves. They boasted a special connection to chimerical beings known as gods. These imaginary friends in sky-high places gave them the right to imperiously help themselves to everybody's share of the common wealth. Even as times changed and they were forced to disown their high-born connection to gods, they couldn't quite bring themselves to own their visceral interconnectedness with the rest of humankind. Royalty of blood gave way to the royalty of money. And while the power structure moved away from the myth that a few people known as kings or priests have a special affinity to gods, its repressive laws and punitive order continue to rest firmly on the myth that a few people known as criminals have a special affinity to demons or brutes. Lawgivers are no longer deified, but lawbreakers are still demonized. One myth down, and one to go!

You have to hand it to them. If a make-believe theology of kinship to gods no longer serves as the foundation of realpolitik, why then, the next best thing is to take the all too real biology of kinship that underlies the brotherhood of man and pooh-pooh it as an unattainable mirage! They can't get away with affirming the divine right of kings any more, but they still get a lot of mileage out of denying our birthright of human kindness.

Why do you think that is? Born as we are with bonding mechanisms, it is with human beings we are meant to bond, not deities. When it comes to the glue that holds society together, a schizoid bond with the gods has outlived its usefulness, but we still haven't gotten around to putting our natural capacity for prosocial bonding to good use.

You have to be deluded in a clinical sense to believe you are some sort of god, set over and above your kind – and yet those who, on the strength of this mental delusion, stratified society for ages and passed themselves off as elites, are not certified as madmen but coronated as monarchs. And what about those who see society as an extension of the neighborly love and communal solidarity that springs from our common humanity? Why, it is customary to this day to dismiss them as kooks and crackpots, cranks and hippies!

We may get a superiority complex from looking down on our fellow man from a height, but if we don't get off our high horse we can never relate to each other as honest-to-goodness human beings. It may be tempting to have such inordinate wealth that we go through life without our feet ever touching the ground, but if we don't climb down from our Pegasus we can never come down to earth, much less live in the ground of our being. We are taught to think of the poetry of love, the song and dance of togetherness, as belonging to an airy, impractical realm. In reality, the hooves of Pegasus have to strike the earth for our rock-hard defenses to shatter and the heartsprings of our humanity to flow.

Astronomy tells us the earth is round. Our psychobiology tells us it is a round table where no one has an edge over anyone else. And then there are those who tell us: *Life is no Arthurian fable. It is a jungle with top dogs and underdogs.* In our heart we know if we got in touch with the ground of our humanity, we would discover a level playing field. And then there are those who tell us: *Get your head out of the clouds and grow up!* We trust we can all happily coexist if we meet our need for nurture, understanding, respect. And then there are those who tell us: *Come down to earth and take your lumps.* And this from the same crowd that for thousands of years justified its dominance of land and sea by hanging on to the coattails of a fantastic horde of imaginary beings in the sky!

What does it say about the eminent people of history that ordinary people cannot fall victim to delusions

THE REPARENTING REVOLUTION

of grandeur, and see themselves as fantastically wealthy, powerful or famous, without being diagnosed as mad? The psych wards are full of people with see themselves as incarnations of gods, as one of the *dii majores* of history, like Napoleon, Christ or Superman. And **we** are the deluded ones for envisioning humanity as one!

And there you have it. Nothing is stronger than our need as humans to bond, to create community, to be part of the larger family of man. But once the power structure can no longer be buttressed by celestial castles in the clouds, or beholden to figments of its own imagination, the only other way for it to bring humanity under its heel is by using our greatest strengths against us. And so out of the abiding need to be thought well of by our kind, we are misled into going along with a system that serves our kind so ill. Out of the abiding need to be accepted by our friends and neighbors, we are tricked into accepting a system that rules us by pitting us against our friends and neighbors. Out of the abiding need to fit in and not be judged as childish or puerile, we are pressured to fall out of alignment with our humanity and lose the very qualities that make us so precious and beautiful as children – our innocence and playfulness, our spontaneity and vulnerability.

Boy, do they have our number! We don't want anyone thinking we're not right in the head for espousing peace and love. We try hard to look sane, sober-minded, sensible. *And what do we get for our trouble? A society effectively run by insane people for insane objectives!*[1] We want nothing more than to live in a world lit up smiles and lightened by laughter of children. For that we are branded a

---

[1] John Lennon, interview (1968)

laughingstock. And then it turns out in order to avoid being made fun of, there is no limit to the misery we are willing to endure!

4. HANGING THE STANDARD
OF HUMANITY UPSIDE DOWN

How do we put traditional stereotypes to the test? Is there something starry-eyed about a society as a communal family whose members mutually support each other and meet dependency needs? Is there something sensible or practical about a society where life is a do-it-yourself project and people can only be babied and coddled at great danger to their moral fiber and hence must be cast adrift to raise themselves by their bootstraps?

No deep political analyses are needed. All we have to do is home in on the relationship between the self and other. In a world at peace and harmony, the relationship is one of bondedness and intimacy. We live and let live by being alive to our feelings. Mutual trust comes from a free exchange of sympathy. Mutual understanding springs from a free flow of empathy. Happiness arises from a lowering barriers on the supply and demand of precious commodities like attention, kindness, support and good will. Altruism is the height of self-interest, since by fulfilling others' needs we naturally fulfill our own.

What is the original template for this relationship? *Our birth experience.* We come into this world to be nestled in close proximity to bodies that nurture and cherish us. Insofar as these caregivers give us room to feel, to live and grow up in tune with our needs and instincts, our mother is the *founding mother* of a state of peace and harmony. Insofar as they respond appropriately to our cries, and make it safe to deepen our attachments and heighten the joys of sharing and caring, our father is the founding father of a civilized state of being. For "civilization, is before all, the will to live in common."[1]

Is this state of symbiosis really pie in the sky? Or is it not the very foundation of our humanness? Does this capacity for bonding and caring make society unstable and impractical? Or is it not at the core of our mental stability as sociable beings? Is our capacity for tenderness and empathy airy-fairy? Or is it not the very essence of the quality that grounds us in our humanity and enables us to realize our potential as human beings?

Then there is another sort of relationship. If you want to call the first sort *lovey–dovey*, it is only fair to call the second *achy-breaky*. For if the first speaks to the integrity of the family of man, the second has *broken home* written all over it. It is marked by a disintegration of bonding mechanisms – a disastrous breakdown in belonging. The self is closed off to the other. It is as if the other does not exist or its existence does not count. There is no shared feeling of support, no mutual rapport. The self single-mindedly pursue its interests with supreme indifference to the happiness and well-being of the other.

Where is the original template for *this* relationship? Once again, our birth experience. But it isn't the one we signed on for. We are not born to be rejected, neglected or abandoned. We don't come to cry ourselves to sleep, far from the lulling beat of another heart, or to peer out forlornly from behind the bars of a crib, never to be picked up and carried. We don't come to be pushed around or picked on, beaten or browbeaten, tortured or terrorized. Insofar as our caregivers force us to suspend the laws of

---

[1] Jose Ortega y Gasset, **The Revolt of the Masses** [p. 76]

our constitution and silence our cries of protest to survive, our mother is the founding mother of a police state. Insofar as we are left high and dry in a place where we don't seem to exist, or our existence doesn't count, our father is the founding father of a state of barbarism. "A man is uncivilized or barbarian in the degree in which he doesn't take others into account."[1]

What about a barbaric tendency to dissociation makes it a model of political sanity? Can the basis of *social* order be a Robinson Crusoe insularity where it's each man for himself? Is this what it means to be *true to life* – to be living a lie, disconnected from our life force? Is this what it means to be a *realist* – to be an unrealized being, out of touch with our authentic self? Is this what it means to be *down to earth* – to be too stunted in our psychological growth to let paradise flourish on earth? Is this what it means to have *both feet on the ground* – to be too emotionally crippled to stand in our ground of being?

How can this topsy-turvy reversal pass itself off as the norm? Feeding our core need for closeness, connection, community – *this* is an exercise in futility, doomed by its sheer impracticality? But the bleak horror of inner fragmentation and outer dissociation, the stark emptiness of discord and desolation – *that* is how humanity is supposed to live?

Such a bizarre inversion of values could never have occurred in the natural course of events. It goes too much against the grain of our heart and soul. Some terrible thing must have caused humanity to flip its core reality on its head. It is a signal of distress – a sign of dire emergency

---

[1] Jose Ortega y Gasset, ***The Revolt of the Masses*** [p. 76]

– to hang Old Glory upside down. So, too, the flagship values that are the glory of our species could only have been turned upside down in response to a catastrophic distress – a dire emergency from which our species has yet to fully recover.

5. THE CRACK-UP OF SPACESHIP EARTH

A few decades ago a plane crashed in the Andes with members of a rugby team aboard. There were 29 survivors. Severe injuries, an avalanche, and the exigencies of high altitude survival, whittled the number to 16. For water they melted snow. Food was a harder nut to crack. Even with strict rationing, food stocks dwindled and soon ran out. There was no vegetation to gather or animals to hunt. And the news on the radio ruled out hope of rescue. The search for survivors had been abandoned. In the grip of maddening hunger they started thinking of eating leather from plane seats and stuffing from cushions. When that proved hard to swallow they succumbed and ate their dead companions. It was not an easy decision by any means. They were making a meal of classmates, colleagues. Only the direst life-and-death emergency could induce them to overcome their instinctual revulsion and visceral horror. Their desperate plight drove them to it. And so a decent, law-abiding group crossed the line and broke one of mankind's greatest taboos.

Here is the key to cracking the mystery of evil. Dire emergencies overcome our kindness instinct and drive us to do things that fly in the face of our humanity. A conscious state is our default setting – but a terrible blow to the head can cause us to lose consciousness. Our conscience is something we naturally live by – but a terrible blow to our survival can cause us to lapse into unconscionable behavior. Armed with a working knowledge of this principle, we can ditch the whole calculus of blame and judgment. Every human being has a limit, which means our capacity to be human has limits beyond which it cannot be overloaded or overtaxed without breaking down. Abnormal acts stem from a breakdown of normality. Inhuman acts demand dehumanizing levels of duress.

We can call this **Mayday** morality. Mayday is a word used by mariners and aviators to signal distress. It comes from a French phrase, *venez m'aider.* In plain English: "*Come help me!*" Mayday is the international cry for help – a cry to be rescued from a life-threatening emergency. If it is answered in time, humans can escape with their lives and dignity intact. It is only when the cry for a helping hand goes unanswered that things can get out hand. Victims are reduced to coping with their plight as best they can. That is when they resort to actions that violate human norms, and may seem unthinkable or unspeakable to those in cushier circumstances. From the standpoint of the victims, they are no evil deeds – there are just last-ditch strategies for preserving the mind and body.

Cannibalism is a perfect catchall metaphor for a wide variety of inhuman behaviors that involve preying on our kind, in one guise or other. History is a litany of abominations that conform, more or less, to cannibalistic patterns of behavior. Rituals of blood sacrifice. Rites of passage through the belly of the monster. A bloodthirsty compulsion to war. Life-sucking practices institutionalized in slavery. The traditional rationale for these gruesome acts of butchery and exploitation is that they are *necessary* to save society from a worse fate. Do they not cry out to be explained by a Mayday mindset? Are they not dire examples of a Mayday Morality that beg to be understood against the backdrop of an earth-shattering emergency, a universal crack-up involving the entire human race?

A month or two after the Andean plane crash, something

serendipitous put it all into perspective. Apollo 17 sent back an iconic picture of the Earth known as the "Blue Marble". It became one of the most widely disseminated images in existence, giving rise to metaphors of our planet as a spaceship carrying all humanity as passengers aboard it.

The sports team aboard the twin turboprop airplane assumed that their passage through a desolate, inhospitable stretch of the Andes would be uneventful. Not in their wildest dreams did they think they would crash land there under circumstances that would lead to unthinkable acts of horror. They were much like passengers on Spaceship Earth in this regard. We take it for granted that our passage through the inhospitable wasteland of space will be an uneventful one – and for centuries it may have been so. But Spaceship Earth has been going around the celestial block for far too long for disastrous accidents not to happen. The odds are that at some point an extraterrestrial object – like say, a comet – ran smack dab into the earth's orbit. *The resulting smash-up would have wiped out most of the human species. Only a handful of victims would have survived under horrific circumstances not unlike those faced by the Andean survivors.*

Ur-survivors went on to cope with their trauma through a range of deranged acts that verged on madness and depravity. As abhorrent as these crimes were, they signaled a Mayday morality. And central to this Mayday morality is the belief that the sins of our ancestors were **responsible** for the cataclysmic Doom visited on them, instead of simply being a coping **response** to it. Though guilt and blame loom large in every Doomsday scenario, that is like saying that God sent Flight 571 careening into an Andean mountain to punish the Old Christian Club rugby team aboard for the abominable sin of cannibalism!

6. SYMPTOMS OF PTSD LEFT
UNTREATED IN PERPETUITY

Flash forward four decades from the Andes Flight Disaster in the fall of 1972. On a spring day in 2012, a meteor exploded high above the Sierra foothills of Northern California with a third of the explosive force of the Hiroshima bomb. The Weather Service reported a sonic boom heard across Northern California and even as far south as Orange County. Fragments were strewn west of the area where gold was first discovered at Sutter's Mill – setting off a modern-day gold rush among geologists and meteorite-hunters.

Flash backwards six decades from the Andean Flight Disaster. On a summer day in 1908, a fireball brighter than the sun, trailing massive dust clouds, exploded over the Siberian taiga. Its force was a thousand times greater than Hiroshima. The column of fire that arose from the earth was 1000 miles wide. Reaching a height of over 12 miles, it could be seen 250 miles away. The blast stripped, blackened and flattened the forest 45 miles from the epicenter, singeing peoples' clothes and knocking them unconscious up to 62 miles away. It blew over men and horses at distances of up to 155 miles. The jolt forced an engine driver on the Trans-Siberian Railway to halt his train 375 miles away.

Both these events have something in common. They occurred in an age of mass communications. The 2012 meteor was a modest one as meteors go. Its impact was confined to the State of California. But newspapers and television, the Internet and cell phones disseminated the news of it all over the world. The impact of the Tunguska meteor was much wider. It was the dawn of the age of mass communication. The newspaper, the telegraph, the radio, and perhaps the telephone could broadcast it all over the world.

But what if a meteor or a comet hit the earth five, ten, fifteen thousand years ago? We're not talking a preindustrial age here. We are talking a prehistoric age, long before books or newspapers – perhaps even before the invention of writing. The magnitude of the event meant that all of humanity was a first-hand witness to it. People wouldn't need big-screen TVs to see it in a wide range of time zones. It was projected on the biggest screen of all – the heavens. They didn't watch it from the safety of their living rooms. They were directly in the path of the celestial fireball, bright enough to be mistaken for a second sun. As it hurtles towards the earth, its long, blowing, glowing tail wraps itself around the planet like the giant plumed serpent Quetzalcoatl. Trees and forests burst into flame. Lakes and rivers boil and evaporate. Rocks glow and shatter in the terrible heat.

Instead of a blast area of a thousand square miles, resembling a butterfly, the damage zone could conceivably resemble a T Rex, well over a half *million* square miles. Tunguska is estimated to be in the 40-100 megaton range. There have been dozens of impacts in the 10-100 range in the past 5000 years...even a few in the 100-1000 megaton range.[1] Can you imagine the scale of devastation? Instead of a few people awoken by a sonic boom you have pandemonium on a universal scale and intensity that is only faintly conveyed by the ritual bullroarers used in the Australian bush. Instead of a few deaths you have die-offs of species. Instead of elusive fragments in the ground, you

---

[1] *The Cosmic Serpent*, Victor Clube and Bill Napier [p. 145]

have craters miles wide, gaseous effluvia spewed into the heavens, drastic changes in sea-level, global tectonic events, magnetic-field reversals, poisoning of the atmosphere and oceans from cyanide in the comet and mass starvation through the prolonged shut-off of sunlight.

We know there were human survivors. A planet of seven billion and counting is proof of that. But what do you think their mental and emotional state would be? Their garden planet lay in ashes, as desolate and hostile to life as the frozen wastes of the Andean flight disaster. The sun is engulfed in darkest night. In the ensuing nuclear winter nothing grows. Despair-stricken survivors are cut off from all hope of rescue because the crash site is not a localized area, but the entire earth itself. There is no way for them to make it except by shedding their inhibitions against cannibalistic behavior and preying on human flesh. *What is the prognosis for a species populated by this handful of survivors?*

The Andean survivors had an entire world that lay unaffected beyond their crash site that could help them recover. In two months they would get to a hospital to be treated for the physical symptoms of their ordeal, like altitude sickness, scurvy, frostbite, broken bones, dehydration and malnutrition – and to a therapy clinic to be treated for flashbacks and nightmares, depression and other symptoms of post-traumatic stress disorder.

Survivors of a world catastrophe could expect no such reprieve. Since no part of the earth remained unscathed, there was no *outside* world that could be counted on to come to their rescue. It would take years for conditions to a semblance of normality. Though the bodies of survivors might heal in time, there were no communal havens to treat their broken hearts and shattered minds. The chances are that while the physical conditions of their trauma would pass, they would retain the behaviors and ideas symptomatic of their post-traumatic stress disorder and faithfully pass them on in perpetuity.

To pass the torch is a phrase used in the context of enlightenment. The primordial cradle of man is a cosmic conflagration – a fiery doomsday. While the biological mechanism for relieving this ur-trauma is available, there are no therapeutic havens for survivors to avail themselves of it. So the torch passed on is a torch of chaos and conflagration, a torch of madness and terror. It is a torch that burns everything it touches. An apocalyptic fire in the heavens lighted a fire in man's belly and made man a devouring fire to man, a purveyor of holocausts, burnt offerings and autos-da-fé through the ages.

Now that we know *violence is the heart and secret soul of the sacred,*[1] it is time to stop beating around the bush, so to speak, and acknowledge that the ritual violence to which religious man returns time and time again is symptomatic of a compulsion to return to the apocalyptic scene of a cosmic crime. Pioneers like Freud and Jung opened our eyes to the compulsive repetition of childhood trauma. Their counterparts like Boulanger in the 18th, Donnelly in the 19th, and Velikovsky in the 20th century paved the way for us to extend the trail of psychopathology from microcosm to macrocosm – from broken home to the breakdown of our cosmic home – from parental violence to the earth-shattering violence of our terrestrial mother and celestial father.

---

[1] Rene Girard, *Violence And The Sacred* [p. 31]

"Terror survives from race to race. The child will dread in perpetuity what frightens his ancestors."

And so yes, return we must to the terror-stricken infancy of our species, but mindfully, not mindlessly this time around, to gain a much-needed understanding of what makes man an *enfant terrible* to man.

7. FROM THE GLACIER OF TEARS
TO THE VALE OF TEARS

Is this all just hypothesis – or is there hard evidence for any of it? There were no newspapers at the time to document the smash-up of Spaceship Earth – *but it is headlined in myths and legends from around the world.* And they all have the same story to tell. There was a time when the cocoon of the biosphere was ripped to shreds, much of life was wiped out, and the burden of saving man from extinction fell on a survivor or two, hard-pressed to make it out alive from the wreckage of worlds in collision. Scholars of religion are the first to point out that many creation stories, including the one in Genesis, "seem to be, in fact, traditions of the reforming of the earth after the Great Annihilation."[1]

Universal chaos, panic, privation – this is the context in which the world of our ancestors had its genesis. As mythic rationales invariably cycle back to what happened in the beginning, all those bizarre, barbaric rituals that revolt our sense of humanity can be traced back to an apocalyptic day of destruction. *Thus our ancestors did in the beginning …thus we do now.* We pass on these gruesome behavior patterns from one generation to the next, religiously replicating them in rituals of murder and mayhem from age to age.

Elements of this pattern may mutate over time, but the thread remains the same. There is a *threat* of social or cosmic chaos. And there is a *promise* of saving ourselves from this dire prospect by preying on our fellow man. The literal form of this predatory behavior is eating the body

---

[1] Mircea Eliade quoting H.B. Alexander, **Myth and Reality** [p. 60]

and drinking the blood of our kind. As a sacrament of salvation it is integrated so seamlessly into our religious tradition we hardly notice it. But the Andean survivors did. When it came time to butcher their comrades and feed on their dead flesh, they salved their consciences, and tempered the horror of their actions, by invoking the image of Holy Communion – the mythic rationale of the son of man who offers his body and blood as food and drink to save the human community from perdition.

What are you saying, someone is bound to sputter in expostulation. *Predatory behavior is a survival mechanism? We turn into monsters from hell to save ourselves from being snatched by the jaws of hell? We have to be eaten alive by self-loathing to make like Jeffrey Dahmer?*

As an excuse for the enormity of evil, we can't let that slide! People don't have to be in the throes of earth-shattering emergencies to be evildoers. They can be sitting pretty on top of the world and still wind up doing godawful things.

Touché! Why do you think the world is in such a mess today? We no longer live under a cloud of doom as our ancestors did. On the contrary, we have what it takes to flood the world with life-easing goodies and peaceful easy feelings. But just look around us. We are locked into a horrific pattern of self-destructive behaviors that still threaten to doom the planet. It's not that the **worst** of circumstances necessarily leaves us unable to put our best foot forward. It's that it leaves an imprint of trauma – and should that be left unresolved, we go about compulsively reenacting it even under the **best** of circumstances!

Look on the bright side. The Andean survivors lapsed into cannibalism but never developed a taste for human flesh. They didn't keep up the atrocious practice in the aftermath of their ordeal. Once it was over, they reverted to a normal diet. *But what if they had never been rescued?* What if they managed to survive on human flesh and had gone on to procreate and multiply in a closed community? What if they enshrined cannibalism as a sacred rite on which their survival depended, and faithfully practiced it to preserve the community? They could wind up as one of those bizarre backwoods communities in horror movies – you know, the kind where carefree road trippers take a wrong turn and stumble on a bunch of inbred hillbillies who hunt people down and eat them for supper.

Doesn't that just about sum up the predicament of the human race? We exalt an eat-or-be-eaten mentality as a sacrament of salvation. In reality, it is a sign we were never saved. *We are all descendants of survivors of a cosmic Andean Crash who were never saved from their doomsday plight.* They never had a lifeline to a sane, stable world. The pandemic scale of doom imprinted them with a Mayday sensibility. And so their survival mechanisms were ritualized in a host of pathological behaviors that in one form or other are practiced to this day. Instead of shaking off as a bad dream the cannibalistic rites they once resorted to as extreme survival mechanisms, we religiously recycle them down the ages until man becomes a wolf incarnate to man. We sacrifice everything kind and gentle and innocent in human nature until dog-eat-dog becomes the gospel of the common man.

The Andean survivors who crash-landed in the *Glacier of Tears* were fortunate enough to walk away and go on to live normal, productive lives. The human race had no such luck. We crashed against an ice ball of a comet, probably repeatedly – and instead of walking away to make a clean break with our catastrophic past, we perpetuated the most destructive aspects of our coping behavior to reduce the world to a *Vale of Tears*.

8. REKINDLING THE HUMAN BRAND
FROM THE ABYSS OF EXTINCTION

Speed is the glory of modern man – though runaway corporate greed does have a way of clogging our urban arteries with its oil-guzzling products and bringing traffic to a standstill. Automobiles are hyped for their power to free us from the physical confines of space – along with trains, planes, rockets. But the greatest form of mobility is psychological, not technological. There's nothing like the freedom of having our emotions for getting up to speed on the glory of our potential. Even the wheel isn't all it is cracked up to be, if it keeps us going round in circles. For centuries the release from the wheel of suffering was believed to come from the extinction of feelings. We now know the only way to prevent the recycling of old trauma is to break down the defenses that deny full access to our feelings. As we create a loving space outside ourselves and use our breath to vacuum up trauma and experientially blow it out of our system, we free up a loving space inside ourselves for joy and radiance, peace and kindness to rush in and flood our beings.

Conversely, we know what kept humanity stuck in a psychological gridlock for ages. Denial and repression. Their immediate roots lie in our individual childhoods, but for our species as a whole, they lie in the cosmic ca- tastrophes that befell it in its collective childhood. Here is something the whole world knew of at the time, yet so few of us remember it today! Obviously, most who lived at the time perished. The tiny fraction that had to repress their trauma to survive condemned future generations to relive it. The news of world catastrophe is archived in the oral traditions of every tribe and region from around the world. *But who today believes the horrific events recorded in myths and legends?* They are encoded as the software of religion – barbaric beliefs and savage rituals that bear the compulsive traces of a period of universal violence. These are the fossilized relics of a time it has been customary to dismiss as ancient history and mythology. But no longer. They are being fleshed out as we speak by the forensic data captured and stored in the hard drive of astronomy, geology and paleontology. Armed with the hard evidence of scientific knowledge, researched have finally become serious about rescuing Doomsday from the murky clutches of myth and superstition. The discovery of *the electric universe* has already sparked a sweeping reevaluation of ancient myths and symbols.[1]

Hitherto our species has been immobilized in the unhealthiest denial of all denials. It has been trapped in a state of amnesia over the traumatic events of its collective childhood. For too long it clung to the fantasy that the cosmos always behaved as a model of decorum, when there were times when it tore out its cometary hair in earth-stomping tantrums. We self-soothed with a Newtonian delusion that the heavens always functioned with the regularity of a cosmic clock, when in reality the earth had its clock cleaned more than once by preternaturally random, senseless outbursts of celestial violence. We comforted ourselves with the idea that stars and planets move with the jeweled precision of a Swiss timepiece, when in reality the human race was terrorized in its formative years by nightmarish

---

[1] David Talbott and Wallace Thornhill, *Thunderbolts of the Gods – A radical reinterpretation of human history and the evolution of the solar system*

rampages that were like something out of an apocalyptic *Clockwork Orange*.

Victims of disastrous miscarriages of parenting can heal. All they have to do is dredge up their tortured history of abuse, in the therapeutic presence of empathic witnesses and understanding advocates, and let go of the toxic energy stored in their bodies and souls. But how can mankind heal from its earth-shattering past if it simply perpetuates the pattern of repression laid down by our ancestors, and continues to exists in a collective state of denial? How can we move forward as a species to embrace the wonderful possibilities of progress if we cannot experientially remember our violent past and so are doomed to recycle it through regressive patterns of behavior right up to the present? How can we empower our young to move forward to seize the joyous possibilities off each brand-new day when we have never really gotten over the trauma of Doomsday?

In South America the Andes flight disaster is known as the *Miracle in the Andes. Miracle* is right. As individuals we are aware we live in a crazy world, and some of us even make miraculous attempts to recover our sanity. As a species we fail to do what we do as individuals for the simple reason we have never faced up to the impact of a time when the norms of planetary behavior fell apart and the cosmos seemed to go ballistic.

Isn't that why we go on mindlessly recycling a destructive message of power and predation instead of mindfully transmitting a procreative message of kindness and co-operation? When are we going to get that tutorial on how generational relationships are like an airline emergency procedure? A mother must affix her own oxygen mask to her nose before helping her children with theirs. Each generation must first secure its own lifeline to the oxygen of love before it can turn around to help the next generation do the same.

Can it be that people still freeze to death on park benches because a medusa of a comet once froze the blood of survivors and left their descendants with a perpetual case of cold-heartedness? Or that hunger still stalks the land because a nuclear winter once blighted the earth's food-growing capabilities and imprinted survivors with a scarcity consciousness? Can it be that the earth's inhabitants are still being incinerated out of the blue by everything from V-2 rockets to drone strikes because our forefathers never recovered from a time when the heavens rained fire and brimstone on the human race? Or that survivors had no way to defend against the terrors of a universe out to annihilate man except by making the fear of annihilation by their fellow man the organizing principle of society? What a horrific possibility! But no less logical than that of a child coping with the horror of an abusive home by growing up to preside over an abusive home of his own.

Why are we in such a godawful hurry to grow up and leave behind our newborn state of innocence, beauty and transparency? We say *we* are the ones we are waiting for. Then why are we in such a mad rush as a society to keep all our appointments except the one we should be clearing our schedules for – our appointment with destiny to show up? Could it be because the only way we can ever make an *appearance* in the sheer magnificence of our humanity is by coming to terms with the impact of the cataclysmic forces that nearly caused our *disappearance* as a species? The universal impetus to come into our own, to shine with the many-splendid potentialities of our being, could very well be the most world-changing force on earth. But we will never know for sure if all we do is go on bringing it to naught, at every twist and turn, by refusing to exorcise the doomsday demons that conspire to return our species to the nihilistic abyss from which it emerged.

9. THE FREEDOM TO GET A LIFE IS THE ONLY FREEDOM WORTH FIGHTING FOR

As much as humanity disguises its beauty, distorts its truth and seems bent on destroying itself, here is a news flash, thousands of years in the coming! *It doesn't have to be like this a moment longer!* We live in a world very different from the one in which our ancestors ritualized a dog-eat-dog mindset. The earth is no longer a flaming wreck from exploding worlds – it is a garden of eden, bursting with promises of peace and plenty. Nature has righted herself and now science has the power to make nature do right by us. The cannibalistic ethos may have allowed mankind to survive under Darwinian conditions of doom and Malthusian gloom. It is no longer fit to survive in this day and age.

Sacrifice, war, slavery, torture…haven't we had enough of institutional bloodlust to last us ten thousand years? Aren't we fed up of the steaming piles of BS we have to step in to worship predatory models of behavior as sacred cows? Is the free market a moloch that can only be appeased by having the economy go through periodic fits of cannibalizing itself? Must we keep our finances in shape through government budgets that serve as procrustean beds to cut off integral parts off the body of humanity? Is there a call for blood sacrifice now that money is the lifeblood of our economy? Must high priests of finance go on compulsively bleeding us dry with debt, stripping us of our livelihoods, ritually whipping up artificial scarcities? Is it to normalize a *dog-eat-dog* world where the way to get ahead is through a cutthroat system of exploitation that does Jeffrey Dahmer proud?

Why do we still act like a bunch of Andean survivors, eating our fellow man for breakfast to survive? At a stage in history when we control the planet, the weather, even space, why is humanity so insecure that it keeps lurching from crisis to crisis? At a time when a cornucopia is spread out at our feet, why are billions of us still living from hand to mouth? Like shell-shocked veterans from interplanetary wars, we have come away with a full-blown case of post-traumatic stress disorder! So what if we are warm and cozy in bed with our loved ones to hold us? There can be no bed-ins for peace as long as we are haunted by paranoid memories and nightmare flashbacks from an apocalyptic time out of mind, and driven to escape by acting out suicidal and homicidal and genocidal fantasies.

Enough is enough! No hulking mountains freeze us out of a world where we can go from surviving to thriving in sixty seconds. The Internet has interknit the four corners of the earth into a public square. The power to touch millions of hearts and minds is at our fingertips with social media. We are no longer isolated and cut off from help. We have the know-how to manifest the most superabundant array of good things in history. We must stop acting like diehard soldiers left behind in remote Pacific jungles, fighting a war of the worlds that was over millennia ago! If we could just break out of our insularity with the breaking news that peace has broken out on earth, we could stop hiding out like troglodytes in caves and abandon the law of the jungle without a backward glance.

Looking forward to a millennium of peace and plenty, we are labeled softheaded for believing in a future that could never be born. Not a word about how the hardheaded sensibility that is the acid test of realpolitik belongs to a past that is long dead and gone. Naysayers scoff at the unreality of a world where our highest hopes and deepest aspirations stand to be realized. *Don't hold your breath for a kinder, gentler world that exists only in our imagination.* And yet their dog-eat-dog agenda passes for sober reality because it is the relic of a world so unhinged by catastrophe that we can barely imagine it.

Remember what the Andean survivors were before becoming man-eaters. *They were a team.* It's time to stand arm in arm and thunderously roar out the answer sages have labored to elicit from us for ages. **We are a team.** Cooperation is how we roll as a species. And there is nothing to stop us from rebuilding that team spirit... that *esprit de corps* that is expressed in the dream of universal camaraderie...*the brotherhood of man.*

For a while we lost that brotherly rapport. We went through a Cronus phase of eating our young, feeding on our kind. *Now we can bring back that loving feeling.* How serendipitous we have a treasure trove of emotional wisdom and intimacy skills to bridge our differences at the same time that we have portals of light to see, hear, and speak to each other across vast distances. Why be trapped in a time warp a dooms *day* longer? Just because our humanity was swallowed by the beast from the abyss, must we go on reimagining it in initiation rites and predatory institutions? We tried saving ourselves like godlike heroes by lengthening our spears and sharpening our swords to rip open the belly of imaginary monsters. Now we deliver ourselves like human beings – by imagining ourselves passing through a loving space, not unlike a maternal belly, to find enlightenment at the end of the tunnel and soak up nurture and support to stake out our place in the sun.

Does this sound at all familiar? *It's time to get over it! Move on! Just grow up already! Feel free to get a life!* Isn't that what we are forever urging each other to do with the horrors we suffered in our individual childhoods? Who knew that where that piece of advice would come in handiest is with the catastrophic horrors we endured in the childhood of the human race!

II. LOVE – THE MOST CONSERVATIVE FORCE ON EARTH

"You know when I said I knew little about love? Well, that wasn't true. I know a lot about love. I've seen it. I've seen centuries and centuries of it. And it was the only thing that made watching your world bearable. All those wars. Pain and lies. Hate. Made me want to turn away and never look down again. But to see the way that mankind loves. I mean, you could search the furthest reaches of the universe and never find anything more beautiful. So yes, I know that love is unconditional."

— Yvaine, **Stardust** (2007)

10. LOVE AS THE ORACULAR
PATH TO SELF-KNOWLEDGE

The world has changed much since caveman days, but not nearly enough in the way we are hoping for. Many thanks to technology for reorganizing the natural environment to fulfill material needs. Where is the science for reordering the social environment to fulfill psychological needs? The seeds of this science are unearthed by dusting off a couple of old Greek sayings. The first is by Archimedes. *Give me a place to stand and I will move the world.* The place to stand or **pou sto** is the stable, unchanging base for changing the world. We need a fulcrum of support that doesn't move to operate the lever that moves heaven and earth. The second saying is inscribed at the Temple of Apollo at Delphi: *Know yourself.* Think of wisdom that rises to the oracular level and this injunction pops into mind. Knowledge is power but self-knowledge is the greatest power of all.

How can these classic truisms reshape the world? What's it even mean to know the self? It means to know ourselves as visceral, instinctual and heart-centered creatures. And only love can shine a light to experientially know ourselves like that. Knowledge is power but first and foremost it is *love.* Our hearts sing with the phrase a pop icon took off his father's headstone.[1] *To know him is to love him.* It could be mankind's epitaph. *The nature of humanity is never so beloved as when we bare it.* How can we even know that when we are indoctrinated from birth to squelch our feelings, shield our vulnerabilities, stifle our truth? Over the

---

[1] Phil Spector, American record producer, songwriter and orchestrator of the "Wall of Sound"

temple of childraising is inscribed the orphic motto of the anti-oracle. *Unknow yourself.* Growing up means knowing the self less and less. It means classifying the integral and authentic parts of the self as secret and unshareable.

If self-knowledge were our guiding principle, history would not be a byword of infamy. Children are open, guileless creatures. In a maternal universe, their spirits romp with expressive and exuberant abandon. All that changes in a predatory monstrosity of a universe. "In the case of nearly all Australian tribes the mothers are convinced that their sons will be killed and eaten by a hostile and mysterious divinity, whose name they do not know, but whose voice they have heard in the terrifying sound of the bull roarers."[1]

Children are initiated into grown-up society by being yanked from the bliss of a maternal universe and thrust into a universe that torments and terrorizes us. It is right to call this universe *hell* as the etymology of the word is to *hide,* to *conceal.* In this hellish environment, we cease to see ourselves clearly – we lose touch with our nature, become strangers to our psychology, shadows of our former selves, hard to understand, inexplicable, invisible. It is a mysterious world of masked men, top secret intelligence, cloak and dagger skullduggery, where we are made over in compliance with the occult standards of a superhuman model, touted by a spiritual and secular power elite. *One becomes truly a man in proportion as one ceases to be a natural man and resembles a Supernatural Being.*[2] It is now that humans begin to act in the senseless, irrational, incomprehensible

---

[1] Mircea Eliade, **Rites And Symbols of Initiation** [p. 8]
[2] Ibid. [p. 132]

ways that are a hallmark of man's inhumanity to man. Our knowledge of fellow humans is shrouded behind the veils that our elders throw over our common sensibility. And not until we tear down these veils can we see them as we are – or see ourselves as they are.

What deters us from reaching out to the bad guys is that they're in such a bad way from never being reached out to. The terrible person is just somebody who needs love.[1] Reading people should take priority over reading books, for all we have to do to love a terrible person is be acquainted with his life story. The heart isn't built to shut out anyone who is an open book. Bookish literacy unlocks the portals of knowledge, emotional literacy opens the floodgates of love. *The bad guys are just good guys whom we never got.*

Some developments are so epochal they change the way we see. The invention of writing is one, the printing press another. Reading and writing are critical to advancing the needle of progress. In fact, illiteracy has long been a stumblingblock to revolutionary change. And now, on the cusp of a new millennium, there is a development more revolutionary than literacy – emotional literacy. *Fluency in the language of feelings* – that could be the greatest paradigm shift of all! It has the power to draw people together in a way that makes peace on earth possible. A poet with his finger on the pulse of humanity, not a president with his finger on the button, dropped a bombshell that wipes out enmity more cleanly than all the armies on earth. "If we could read the secret history of our enemies, we should find in

---

[1] "Everything terrible is something that needs our love." Rainer Maria Rilke, German lyric poet

each man's life sorrow and suffering enough to disarm all hostility."[1]

When it comes to traditionally cracking open human beings to the truth, is there much difference between past and present? In the past, the *true man* emerges through initiatory tortures that involve stress positions, sensory overload, sensory deprivation, forcible extraction of teeth, hair, nails, etc. In the present, the *truth in man* is extracted through torturous rites like electroshock and waterboarding. What if the *science of coercion* as practiced in psych labs and detention centers is doomed to go the way of the initiation rite? What if the future belongs to a *science of love* as practiced in personal growth workshops and self-awareness retreats? It is here that groups of strangers gather for impromptu readings from the secret history of their lives. And here that the equation of love and knowledge hits us full on in the gut. It matters not how closed-off or clammed-up we are when we walk into these bare-and-share sanctuaries of the soul. By the time the defenses slip, the disguises wear out, and the evasive maneuvers are given a rest, we show up in each other's presence in the secure conviction of standing in the ground of love. *When people really get to know each other, in detail, they fall in love with each other.*[2]

In torture chambers and black sites our humanity is broken and twisted beyond recognition. Only in havens of non-judgment does our truth come out and our true selves surface in all their magnificent wholeness. We don't get people to spill their guts by violating the Geneva

---

[1] Henry Wadsworth Longfellow, **Driftwood**, 1857

[2] Brad Blanton, **Practicing Radical Honesty** [p. 198]

Convention. We do it by honoring to UN Convention on the Rights of the Child. We don't use duress to uncover terror plots. We use love to expose the litany of pent-up hurts that drive people to be holy terrors. We don't use aggressive interrogations to extract scraps of evidence to confirm a ruptured mindset of *us against them*. We use asylums of love to share our universal heartbreak and confirm our common humanity.

11. THE POWER OF PSYCHOLOGY TO SHRINK THE WORLD

Thus the Delphi oracle only had it half right. It told us to *know the self* but left out the best part – to *know, know, know the self is the same as to love, love, love the self.*

And Archimedes? Science has found a *pou sto* in the ground of nature for levering the world. But overcoming physical barriers is nowhere near as challenging as overcoming psychological ones. What blocks our progress now are not mountains and oceans but the emotional walls and defenses we put up to each other. The real catalyst of world change is not freedom of physical motion but freedom of human emotion. A *pou sto* in the ground of nature moves our bodies through space. A *pou sto* in the ground of *human* nature moves our souls through a timeline of feelings. Technology enables us to hate and obliterate over vast distances. Only a science of human relationships enables us to bridge our differences and come together as one big happy family. It is not for nothing that psychotherapists are dubbed shrinks. Technological change shrinks the world to the dimensions of a global village. Therapeutic change shrinks the judgments and prejudices and fears of intimacy that keeps us apart and ensures that our global village is not some blood-curdling village like Covington[1] but a heartwarming small town like Mayberry.

You know the phrase *down-to-earth*? Being in close proximity to our planet is a metaphor for being practical, sensible, realistic, with no illusions or pretensions. But there is something more personal than the earth – our

---

[1] The setting for the psychological thriller film, **The Village** (2004)

71

humanity. How come we don't say *down-to-humanity*? And if we did what would we mean by it? Being down to our bones, our viscera, our lungs, our heart? Being *down to planet humanity* is not a metaphor for being in touch with ourselves. It is the real deal of being in touch with our human essence – our needs, instincts, feelings. Our bodies support and sustain us as our mother earth does. From the instant of conception we live in a maternal womb. When born we are laid on our mother's breast, not the breast of mother earth. As an external fetus we are carried by human mothers. As infants we need to have our heads supported. As adults we hold our heads up high, but never outgrow the need to have our hearts uplifted.

As we are supported by others, we in turn support others. For a mutual-support society to flourish, it is not enough for our feet to be planted on the ground. Our spirits have to be centered in the ground of our being. *What miracles could we not accomplish by staking out that holy ground?* We could be moved to our depths to feel our pain and thereby be moved to feel the pain of others. We could circulate the currency of unconditional love to crash a banking system that circulates money as debt. We could disarm all hostility, disband all armies, dispose of all arms as museum relics. If our hearts could bleed for our fellow humans as creatures of flesh and blood, we would outgrow our infatuation with gods and wannabe gods. Religion would be drained of fratricidal sound and fury. Mass media would be flooded with the sunlight of truth and disinfected of propaganda. And the machinery of progress would lever earth into alignment with heaven.

In a word, the self-knowledge that comes of being grounded in our humanity is inseparable from self-love.

Self-knowledge and self-love operate in a feedback loop. To love ourselves as we are we first have to know ourselves well enough to establish who we are. And we can't really know, still less be true to ourselves, without escaping the maelstrom of pressures that drives us to be *beside* ourselves with anger and hurt, and relaxing in a space that is loving enough to make us feel accepted, appreciated and understood.

Life is a struggle because we strive to live contrary to our nature. It is only when we live according to the lights of our nature that we experience the ineffable lightness of being. Life is hard because we have grown hardened to our feelings and desensitized to the point of becoming hard of hearing to one another's cries. We have to soften our hearts to experience the soft life. We have to be good to each other to have the good life. By reviving our capacity to feel we come superabundantly alive, ebulliently real, brimming with spontaneity and verve, bursting with curiosity and creativity. It is from this place of groundedness in the heart and soul of our being that we wield the master lever that makes light work of all the heavy lifting required to pull off the brotherhood of man.

The little Scottish girl has already showed us the way. When she was seen carrying a big baby boy, someone remarked how heavy he must be. With surprise she replied, "No, he's not heavy; he's my brother."[1] As for the rest of us, it is only when we stand in the *pou sto* or ground of our humanity that we can really, truly mean it when we say of every single person on this planet: *His welfare is of my concern/ No burden is he to bear/ We'll get there/ For*

---

[1] As related by James Wells in his book, **The Parables of Jesus** [1884]

I know/ He would not encumber me/ He ain't heavy, he's my brother.[1]

---

[1] *He ain't Heavy, He's my Brother*, a worldwide hit ballad performed by The Hollies

12. PUTTING THE WORLD TOGETHER
– THE FLIP SIDE OF E=MC²

The graphite at one end of a pencil is for writing, the pink rubber at the other for erasing. Perhaps the urge to erase or wipe out is not an unqualified evil. Perhaps we destroy in an attempt to tear down things that are built up all wrong and out of whack.

A cosmic fist once smashed into earth and knocked mankind unconscious. When mankind came to, it was too woozy to remember who it was. It set about remaking itself according to a woo-woo construct, not so much conceived in sin as out of sync with human nature. When a 12th century Persian philosopher, mathematician and poet dreamt of starting over, it was naturally by raising a giant fist to smash the world again, so this time it could be remade according to a heart-centered model more user-friendly to humankind. *Ah Love! could you and I with Him conspire/ To grasp this sorry Scheme of Things entire,/ Would not we shatter it to bits – and then/ Remold it nearer to the Heart's Desire!"*

The 73rd quatrain of the Rubaiyat is a clarion call for change that galvanizes the spine and cause it to pulse and undulate with anticipation! But poetic license aside, there are practical things to consider. Before running out to smash the world, we might pause to ponder what *is* our heart's desire? The heart is our center of emotional intelligence. There must be a certain logic in choosing our partner in crime when it comes to changing the world. We don't say, *O God, could you and I conspire.* Nor do we say,

Ah Caesar, or Ah, Croesus, could you and I conspire. We blurt out, *Ah, Love could you and I conspire!*

After smashing the world, we don't recreate it in the image of a divine Creator, much less from an abstract Cartesian diagram in our heads. We recreate it according to our *heart's desire.* Why not? Who would want to smash the world anyway unless it continually frustrated the heart's desire? In case you never noticed, ours is not a heart-centered world. If anything, its miseries and privations break the heart. The world of our desires is a place beyond heartbreak – a place where the broken pieces of the heart are mended and made whole again, and where that wholeness is so akin to holiness that we would sooner smash the world than break a single heart. Then again, if we came up with a world designed to heal the heart and preserve it whole, how could this world of ours manage to keep up its heartless facade? More likely than not, *it* would shatter to bits!

So when the moment comes to move the world from the *pou sto* of our being, we don't waste time puzzling from where to where exactly. Through the eyes of the heart, the world is a broken place. It is conceived in the image of universal chaos and bears the imprint of apocalyptic doom. As heirs of a race of ur-survivors we don't have to wonder, *what would the contours of my being be like, if reality had taken the form of a universe that was out to annihilate every trace of me?*[1] All we have to do is look around to see the many distorted shapes humanity has taken over time.

---

[1] Toby Mostysser, "The Weight of the Past, Reminiscences of a Child of Survivors," **Living After The Holocaust**, edited by Lucy Y. Steinitz with David M. Szonyi [p. 20]

From the instant we draw our first breath we are bombarded by *Don't Be* messages. And until the instant we breathe our last we are brainwashed by fears that the world is a dangerous place out to get us. And it shows, doesn't it, in our chronic deficiency of *joie de vivre* and the scarcity of a commodity like *relaxed joyousness*. The world is always ready to grab us by the throat. It never tires of playing games of *gotcha*. It is forever putting the squeeze on us. The superstars of our firmament keep landing on our heads and sucking the oxygen out of our environment. The roof is never not threatening to fall in. The fear of being judged and found wanting continually drives us underground to hide. The lines of our existence are drawn tighter and tighter around us, stressing us out, distressing us, reducing us to dire straits and constricting our ability to grow, travel, expand our horizons, live large, and breathe.

That's our world as it now stands. And where would we move it to? To a place where the human race is free to let down its hair at last, because it is no longer haunted by the specter of a doomsday comet brandishing its horrid tresses across the sky. To a place where we can have a beautiful day without worrying that it is our last. To a place where we know what it is to be embraced in the darkest, most shame-filled corners of our being – a place where our souls no longer have to cower out of sight or hearing, because we have what we have always wanted from life, which is to feel ourselves beloved on earth.[1]

And there it is. That pivotal shift from a world ***out to annihilate us*** to one where we are ***beloved on earth***. It can only happen by mastering the experiential knowledge of

---

[1] Raymond Carver, *A New Path to the Waterfall* (1989)

ourselves and taking a definitive stand in the ground of our being. *Pou sto* and *gnothi seauton* – a place to stand, a self to know – are two lustrous pearls of wisdom in their own right. Combine them and you have the breakthrough formula for world change. By knowing ourselves, we discover that *irreducible* part of us that is the same for people everywhere. And by taking a stand in the ground of our being, we have a *pou sto* from which we can move the world in a direction that does justice to our common humanity.

The union of the emotional intelligence of Delphi and the mechanical insight of Archimedes gives birth to an earth-shaking formula that is the spiritual corollary of $E=mc^2$, and at least as powerful, if not more so. $E=mc^2$ simply unleashes the destructive energy of the cosmological dimension to grasp this sorry scheme of things and shatter it to atoms. It is the union of *pou sto* and *gnothi seauton* that releases the creative energy of the psychological dimension to put back the pieces of our broken heart, so that we have the original to go on when it comes time to remake the world closer to our heart's desire.

13. BUILDING SOCIETY'S FOUNDATIONS IN THE AIR

As inhabitants of earth we automatically have an affinity for things earthy and solid. The terrestrial ground under our feet is home base. When we stand on the earth and look up at the sky, we see mostly empty space. Our greatest fear is having nothing under us, nothing to hold us up or support us. Our deepest horror is stimulated by that which appears to be bottomless – the *abyss* that has no *bottom* (*a*, without, *byssos*, bottom). A grand shout-out to gravity for saving us from endlessly falling through the immensity of space or floating upside down in the heavens without any prospect of ever coming to rest!

When we want to discover the truth we say we must get to the bottom of things. By implication, that which lacks a bottom is untrustworthy or untrue, confusing or corrupt. Those who lack truth or principles gravitate to places that lack a bottom. Which is why the bottomless pit is traditionally reserved for the devil. And why we must have a biological place to stand, and a psychological self to understand, to stake out the high moral ground. As a species we keep backsliding into the abyss because our world lacks an organic foundation, a firm footing in the ground of our being, and our lack of self-understanding and self-knowledge verges on the abysmal. As rapidly as electrification programs have progressed in urban and rural areas in the past century, whole areas of the human psyche remain off the grid, shrouded in mystery, and clouded with ignorance.

A new window on this subject opens up by highlighting the word *real*. Our affinity for the real comes from a planet filled with things (*res,* thing), as opposed to an interplanetary space which is a void. Everybody wants to be real. Iterations like *for real, get real, the real deal* pepper our everyday speech. We gravitate to the real because we want something to get our hands on, or plant our feet on, as opposed to a phantom or mirage. *Reality is the real estate of the mind.* We believe we are better off for owning a piece of it. And how do we lay title to *reality* from an everyday perspective? The commonest way is through a spatial orientation. The real is associated with the bottom end of a vertical axis. Things are said to have a *foundation* in reality. Phrases like *down-to-earth* describe those who operate from a sensible place. Pragmatists have *both feet on the ground.*

Conversely, the realm of unreality occupies the top end of a vertical axis. The unreal is said to be *baseless* or *groundless.* The fanciful or visionary is called *pie in the sky.* Those who lose touch with reality have their *head in the clouds* or live in a *castle in the clouds.* They are derided as *spacey* and *airy-fairy* and dismissed as natives of *la-la land.*

In light of this orientation, you would think that when it come to organizing society, the most real or reliable input would come from the ground – from those closest to the foundation or base of things. Whereas that which comes from above, from the high places out there, would be distrusted as illusory or unreal. *Boy, would you be mistaken!*

Through a Mayday inversion, reality is mapped out by those at the upper end of a vertical axis. Those who dwell in castles in the clouds are the same people who build castles of stone to dominate the landscape. Apparently liberated from the down-to-earth constraints of ordinary mortals, they tower over society with every intention of lording over it. Laws are devised and enforced by those at the top of the pyramid. In this Bizarro universe, reality is structured from the *top down*, not the *bottom up*. Ironically, the high muck-a-mucks who give themselves airs are contemptuous of the ground that supports them. It may be the very basis of their existence but they consider it to be *beneath* them!

How can this *top-down* model of rule bode well for humanity? An elite of castle-dwellers in the clouds cannot help but be singularly out of touch with the needs of ordinary people. They are no longer addressed with honorifics like Pooh-bah or Pharaoh, but staffs of white-gloved servants, public and private, still hover around to do their every bidding. Dream teams of advisors and lobbyists scurry about looking out for their every interest. Battalions of attorneys and security personnel are ready to take a bullet to save them from the consequences of their actions. An extensive portfolio of resources exists to buttress their cushy existence and insulate them from the demands of everyday reality.

Not all bubbles are created equal to preserve the status quo. Those who want to live peacefully off the earth's bounty are accused of living in a bubble that must be regularly burst. Those who function as a giant energy suck on mankind live in the biggest of bubbles, and none dare burst it without being branded an enemy of the state! The Ten Commandments of politics all boil down to one: *Thou shalt not burst the bubble in which the wealthy and powerful live.* So this bubble is guarded night and day by firebreathing dragons whose sole function is to keep it inflated with an inexhaustible supply of hot air.

Needless to say, the few cannot live *large* in palaces and castles without consigning the many to live *small* in sculleries and stables, salt mines and sweatshops. *The grim reality mankind resigns itself to is shaped, not by the actual rigors of the physical world, but by the arrogant demands of self-styled emissaries of a supra-natural world.* A mass effort to maintain a class of wannabe gods in the splendiferous excess to which it says it is entitled drags everybody else down to live lives of drudgery and degradation. Ours is the lot of the *Man with the Hoe*, bent double under the onus of balancing our checkbooks, and ground underfoot like a stolid beast of burden. Theirs is the lot of the *Magician with a Wand*, conjuring money out of thin air to inherit the earth and sit on top of the world.

In our world the rockbottom certainties are love and family; in theirs, death and taxes. As if taxes are as unavoidable as death, when in fact they are a cheap confidence trick for confiscating vast amounts of our common wealth to keep an 'elite' in the high-flying lifestyle to which it is accustomed. We intuitively grasp that the heads of the high and mighty are in the clouds, their home is the airy-fairy realms of gods. Yet we dutifully fall in behind them as *authority figures* and think nothing of letting them push us around and run us ragged, even though death itself could be easily avoided in the form of the catastrophic death toll we suffer *en masse* in every century by humoring their mania for war.

Who are these higher-ups in their command-and-control headquarters who tax us to death to pay for the cost of mobilizing and marching us off to our doom? They are the **monarchs** whose right to rule flesh-and-blood mortals derives from disembodied bloodless beings who cannot even attest to their own existence, much less have any claim on ours aside from the faith we repose in them! They are the **money power** whose right to impose a fiat rule on mankind derives from bits of paper with no intrinsic value or worth aside from the trust we repose in them! Is this what the whole basis of civilization comes down to? A faith in *holy scrip* that is objectively as baseless as a faith in *holy scripture*?

14. DISMISSING THE FOUNDATION OF HUMAN NATURE AS AIRY-FAIRY

It seems our species is left spatially challenged by the impact of cosmic disasters. Humanity no longer knows which end is up anymore. It is bad enough the foundations of the power structure are effectively suspended in midair, like some nonsensical construct out of Lewis Carroll or Norton Juster.[1] The enormous toll on our material resources to keep this elite class walking on air with its head in the clouds is just the beginning. It doesn't cover the sky-high emotional cost to our humanity to subsidize this upside-down arrangement. For if the upper end of a vertical axis is what dictates our reality as a *civilization*, what becomes of the *lower* end that defines our biological reality as a *species*?

You guessed it. It turns into a source of *unreality*! The primal need for bonding and community is treated as an airy-fairy thing! Those who look to gear society to fulfill our prosocial needs are said to have their head in the clouds! What a mind-boggling turn of events! The power structure that allows a tiny minority to nurse delusions of grandeur and make hubris the state religion is enshrined as the touchstone of reality. And the love that is the bedrock of humanity – the touchy-feely connections that ground us in a state of mutual interconnectedness – is written off as stuff we have to *imagine* and *dream* about!

This preoccupation with the top as the headspring of reality has a predictable effect. It cuts us off from the

---

[1] "...For standing directly in front of him (if you can use the word 'standing' for anyone suspended in mid-air)..." Norton Juster, *The Phantom Tollbooth* [p. 102], Special 35th Anniversary Edition

heartsprings of our humanity and leaves us high and dry. Our experiential reality as mammalian organisms loses a great deal of its living authority. Our emotional truth, instead of taking center stage in the forefront of consciousness, takes a back seat, out of mind and out of sight, becoming the last thing we tend to focus on or take into account. Our natural web of needs and instincts, instead of serving as the bottom line in our dealings with one another, is now consigned to the bottom of the heap!

Through a cockeyed fixation with *heads* of organizations and *heads* of state, we turn our backs on the *heart* of humanity – our true seat of higher learning, our center of emotional intelligence. We give pride of place to the dead letter of a constitution framed a few generations ago by an assembly of men of wealth and letters to benefit their economic class. And the most seminal constitution of all – our living constitution, framed over millions of years of evolution to benefit the human race – is treated as if it were some sensitive, personal document to be put through the shredder to keep it from falling into the wrong hands. We don't even have to study its articles in dusty law libraries. We breathe, eat and sleep its system of ruling principles. Its body of integral laws is one we inhabit for life. And all we have to do is be governed by its laws to function as sovereign beings in our own right, emotionally literate and entitled to live rich and rewarding lives.

How then can a blind faith in the authority of sovereigns come naturally to humans who are sovereign beings in their own right? People look up to authority figures because they are taught to downplay their role as the creators and originators of social order. Consider a mother who is there to celebrate her baby's first step or first word, who pushes her daughter on a swing or reads her a bedtime story. Or a father who fixes a snack for his son, sits down to talk about his day, or helps him to craft a model airplane. All the 'little' things parents do to communicate love and caring – the back rubs, the board games, the shared jokes, the *I love you* notes in lunch boxes – seem ordinary but in fact are extraordinary in their contribution to world peace and security. These little acts give children their sense of dignity and worth. They give their world meaning and order. Out of these everyday demonstrations of caring and nurture comes the trust and well-being that stabilizes our world and keeps it safe. A terrorist who wreaks havoc on the cozy, idyllic world we take for granted is simply someone who had no part of it growing up.

Those in the vanguard of creating a stable, secure world for children are not princes and plutocrats. They are parents and educators. From prenatal vitamins and baby slings to courses in non-violent communication, the parenting complex is on the cutting edge of establishing relationships that enrich the lives of children and make them deeper and more meaningful. Whether one realizes it or not, this is the cornerstone of a peaceful democratic world. But the people whose everyday efforts make this possible are never really valued for their contributions to the social order. The are relegated to the bottom end of the social hierarchy – and for the most part are underpaid, or paid not at all. They are taken for granted as part of the background of society, while the attention and kudos goes to the 'big boys' with a whole other toolbox and skillset to reshape the social order.

Do we have any idea what a terrible price we pay for

this arrangement? Just think if the trajectory of birth and parenting defined the arc of our moral universe. We would not only bring the bodies of the next generation into the world but grow their souls to *appear* in the magnificence of their humanity. And then what need to invest in systems of policing and armed confrontation that aim to *disappear* people from the face of the earth? If we spent more time and energy giving children the support they need to let their light shine as open-minded, open-hearted beings, what need to waste money and manpower on regimented organizations that lock them up in prison cells or zip them up in body bags?

It is no coincidence that our regard for love and caring is warm and fuzzy, light and fluffy, while our rage for death and destruction is hot and heavy-duty and industrial. We have all kinds of excuses for withholding what children *need* in the way of kindness and attention, while gung-ho to give what we assume they are *asking* for in the way of violence and aggression! How are the promises of fathers so iffy and unreliable, stringing children along through years of anguish and disappointment, but when children grow up to take out their frustration on society, the punishment meted out is so swift and sure?

Hasn't it crossed our mind that if touchy-feely ways of relating were not judged so airy-fairy, it wouldn't be necessary to crack down on our fellow man so heavy-handedly? Don't we know that if family members made a daily practice of exchanging heartwarming glances, we wouldn't be living in a world where it is normal to exchange gunfire? How could kids held aloft in outstretched arms by parents, who beam up at them with joy in their hearts, ever grow up to be spreadeagled and handcuffed, face down on the pavement, to be looked down on with disdain by those entrusted to protect and serve?

There is only one way to glorify the work of soldiers who maim the able-bodied and wound and waste them by the millions – and that is by demeaning the work of mothers who rejoice at bringing into the world babies with ten fingers and ten toes, cherishing their wholeness at every step to help them reach the fullness of their potential. If people could see their way clear to picking up, holding and comforting *crying babies,* we would have no need to dedicate stretches of highway to the exploits of the *Screaming Eagles.*[1]

When are we going to make the connection? In a society that treats the mothering that protects children and keeps them safe as a dirt-cheap commodity, the only way we can be safe is through a militarism that costs an arm and a leg. In a culture where the power of love to create good people is treated as a sweet airy nothing, the cost of using brute force to eliminate bad people becomes a bitter pill, a crushing burden on society!

The only thing more pervasive than our culture's enthusiasm for the *bang-bang boom-boom* method of problem-solving is its contempt for the *touchy-feely lovey-dovey* approach to social relations. One of these definitely merits derision but it is not the tactile, emotional interactions that are so crucial to the experience of love. When we realize the need for love is a non-negotiable thing, force becomes the most unnecessary thing in the world. When we submit

---

[1] "Freeway Stretch Honors 101st Airborne" (an Army Division famed for its assaults in WWII, Vietnam and Iraq.) Gregory W. Griggs, *Los Angeles Times,* September 22, 2005

to attachment parenting and social bonding as the first law of our nature, the practice of isolating lawbreakers in prison cells loses all credibility. If we took the power of love more seriously, a culture of militarism would be beyond ludicrous and grotesque! And if lovey-dovey touchy-feely interactions ever got the reverence they deserve, the whole bang-bang boom-boom mindset that actuates our foreign policy and a majority of film plots would be a blasphemous joke of literally cosmic proportions.

15. POWER AS THE SEMINAL SUPERSTITION

The high and mighty give fresh meaning to the word *superstition* – literally, the act of standing *over* others instead of standing *by* them or *beside* them. Power is the seminal superstition – a system of misbelief that runs on ignorance and fear and violates the norms of common sense and human sensibility. The powerful are untethered from the ground rules of humanity. Chains of indentured servitude seem more practical to them than bonds of fellow feeling. They are consumed by irrational hungers unplugged from legitimate needs. Detached from ordinary folk, and remote from the give-and-take of human existence, they are natives of la-la land whose patriotic loyalty is not to any earthly country but their place of origin. They begin by going off the deep end and end up going into deep space in a quest to colonize earth like an alien planet. They struggle to sate their lust for power and opulence by controlling the world like a mad scientist, but fail to satisfy the heart's capacity for joy and peace through a science of human relationships.

And so what becomes of our need for authentic contact and coexistence? What becomes of the messy, cluttered world of clapboard houses where most of us live, with ice cream trucks chiming *Für Elise* on the streets, food stains on the linoleum and tricycles in the driveway? What becomes of the world where truth rings out from the joyous squeals of scampering children, playing tag with jets of water that randomly shoot up from the ground? It is drowned out by the agony of grown-ups who squeal to avoid being waterboarded by those on a witch-hunt to give credence to their paranoid fantasies.

Is it progress to surmount a mountain of scruples, or streamline morality to stop at nothing? Is the measure of civilization not all the things it values ahead of money, but how much it values money before all things? Should our moral decisions be subject to time-motion studies, so that all that laborious hand-wringing that goes into dropping a bombshell that causes a single person pain, is replaced by an instantaneous snap of the fingers when it comes to dropping bombs that incinerate tens of thousands of people?[1]

Who do you think sets the bar of heroism higher – a woman who gives birth without drugs or a man who kills in an anesthetized daze? What demands a higher form of courage – hiding one's feelings behind the conventions of the strong silent type or stepping out of one's comfort zone to engage in raw honesty and emotional risk-taking? Does closure come from kicking down doors in search of insurgents – or breaking through barriers in search of intimacy? Pseudo-heroes spills the blood and brains of others without ever examining what makes their lives worth living. Honest-to-goodness heroes spill their guts to get to the heart of the relationship with their fellow human beings.

The sweet spot of humanity has always been the place where we umbilically connected to our maternal universe. But to nourish our humanity from that spot is passed off as a self-indulgent form of navel-gazing, a namby-pamby distraction from the business of real life. And the only place left to luxuriate in the epiphanies of beauty and wisdom that flow from our hearts and souls is the poetry of pop songs heard over the radio, or through streaming audio,

---

[1] "When an interviewer asked President Truman whether the decision to use the atom bomb was a morally difficult to make, he shot back, "Hell no, I made it like that," snapping his fingers." japanfocus. org

and the feel-good screenplays scripted by Hollywood dream factories.

As long as we are directionally confused as a species, the superstition of power has us in its benighted grip. The *top-down* orientation to which we give credence is not just a spatial affair. It is an entire value system. Its status and preeminence, its lofty dignity and worth entitles it to organize society and lay the foundation of political rule. Stability, law and order – these are values drilled into us from the cradle. They are all bound up with a top-down hierarchy and its authoritative dictates. A stable society ostensibly depends on an authoritarian power to preserve law and order imposed from on high.

Hang on a minute. If we are to talk *stability*, what is more rock-solid than the earth under our feet? What offers greater firmness than the ground of our being? Isn't that where we are set, fixed and settled in our nature? How can some top-down structure that may as well be standing on air, standing as it does *over, above* and *beyond* the constitutional limits of our humanity, ever serve as the foundation of human stability and order?

By the same token, if the ground of our humanity is like mother earth under our feet, why are we not standing firm on it? The answer lies in a simple word for a complex experience. *Love.* Love anchors us in our maternal ground of being. It means the world to us even as it gives our world meaning. Through love we are in touch with reality. We enjoy the support we need to be true to ourselves. Love familiarizes us with the inner workings of human nature. It is the laboratory for the science of human relationships.

Conversely, what earthquake causes the ground of our being to heave and shake, rattle and roll? What tectonic shift causes this ground to split open and swallow us up. A

breakdown of love. Without love we lose our foothold in our ground of being. The bottom threatens to fall out of our lives and we are in danger of falling into a bottomless pit.

Given that love is the mainstay of peace and order in our homes, you wonder why it is not the guiding principle of legislative houses tasked with keeping peace and order in the land. It must be because the word on the street is that love is airy-fairy. You can't live on *love and fresh air* – as if love itself is not the oxygen of our lives and our lungs need something grittier, denser or dirtier to clog them up and keep them going. And that says it all. Forget about all the cliches on how love hurts. *The only thing wrong with love is that it has a PR problem.* No matter how much it is celebrated in Motown ballads or medieval romances, the impression we are left with is that it is a diet of sweet nothings. Silly love songs are stitched with rhymes like *moon, spoon* and *June,* but only romantic fools believe that love is the rhyme and reason of our lives. Sensible people know that life demands that we be made of sterner stuff. The sooner we quit wrapping our arms around each other, and start wrapping our heads around realistic, matter-of-fact, practical concerns – like good grades, college diplomas, money and jobs – the better off we will be.

What will it take to change our bias against love? Must we wait for some Isaac Newton to prove that love is the interpersonal force of gravity that holds our species together? Can't we just listen to Wayne Newton breezily give thanks to the gravitational attraction of love across time and space? Can't we look into our hearts to discover the laws of emotion that keep our bodies revolving around each other? Luckily, we've never had to experience how, without physical gravity, we would fly off into space. But history books and police blotters are crammed with examples of how, without the gravity of love to bind us to each other, we lose our grounding in our humanity and fly off the deep end.

16. DON'T GO CHANGING HUMAN NATURE

It is too ironic for words that love, especially when coupled with romance, is fodder for the realm of imagination and idealization – a fanciful, quixotic sentiment that is fun while it lasts, like some escapist idyll or honeymoon. Love may be equated with heartfelt poetry, and power with *realpolitik*. But there is a reason why the poetry of love is broadcast from our radios all day long, while the voice of realpolitik never carries beyond soundproof backrooms and electronically shielded cones of silence. Most people naturally revel in the one – and are morally repulsed by the other. And it is not because *they* lack wisdom or morality, tolerance or responsibility or other markers of a developed humanity. It is because their *leaders*, cut loose from their moorings in the ground of humanity, are raring to go to any immoral, inhuman lengths, as long as they are not caught.

How can rationalists mock the touchy-feely power of love, when without it their brains wouldn't have developed as infants? How can realists superciliously look down on the touchy-feely power of love, when without it they would be vegetating in a semi-autistic state like a Romanian orphan? *Without love, where would you be now*? Why is that just the refrain of rock musicians when it should be the mantrum of every lawmaker, opinion-maker and policy-maker in the land? Why is it just a song by the Doobie Brothers when it should be the rallying cry of a mass movement for the brotherhood of man?

Social order is not a function of political power. It is the power of touch that destresses and calms us, centers and stimulates us with the vital contact we need to turn an encounter into a prosocial event. It is the power of feelings that releases us from our demons, breaking the spell that turns our days into nights on Bald Mountain, and setting our hearts pealing like church bells to herald our return to a place of sweetness and light.

For thousands of years governments smugly wrapped themselves in a mantle of conservatism. But what *is* it? An upstart system based on parental force and authority in the private arena, and political force and authority in the public arena? Or a system rooted in biological instincts and imperatives, adapted over millions of years of evolution?

Where are we going to find that venerable tradition as old as the hills? In a new-fangled system of customs and rituals, as bizarre and irrational as they come, that must be shoved down the throat of every generation so that it tamely submits to symbolic or real forms of mutilation and self-sacrifice, dismemberment and death? Or an ancient system of biological mandates and expectations the young of our species bring into the world from time immemorial to foster social bonding and the convivial boons of community?

What is the touchstone of the real and true? Is it an extrinsic set of religious laws from the gods know where – or a body of laws intrinsic to the human organism? Is it political customs befitting those who play god over man – or behaviors encoded in the ancient continuum of our species, custom-made to the needs and expectations we evolved with?[1] Who are we kidding? The answer's a no-brainer. Nothing is better suited to meet our evolutionary needs and expectations than a dispensation of love. Our lungs

---

[1] Jean Liedloff, *The Continuum Concept* [pp. 22-23]

are made to be filled with air, our eyes to be filled with light and our hearts to be filled with love.

If we retain our humane sensibility, our sensitivity, our very sanity as humans, it is through the experience of knowing what it is to be loved. And if there is a long-established order to humanity that is the one and only *social* order that passes every test of conservatism, it is through the good offices of love that we are faithfully married to it.

Love as such is the most conservative force on the planet. The capacity to love and be loved is the *bedrock* of human nature, speaking to the irreducible part of us that is changeless. We cannot do honor to the establishment of this bedrock principle without becoming *de facto* conservatives! Indeed, if conservatism has any real meaning, it is precisely this. When we are grounded in our established nature, when we are committed to promote and preserve the interests that flow from the heart of our being, we are conservatives in the deepest sense where it counts. Not enough of us are **conservatives** in this authentic sense. If we want to move forward as a species and change the world for real, we have to align the world with that part of our humanity that is conservative to the core!

As counterintuitive as it seems, this **conservatism** is the heart of real change. As much as we gear change to progressive or liberal values, they seem not to stick in the long run for a reason. *We fail to source them to the conservative core of our being.* Human nature is the ultimate repository of conservative values. There is no need to stretch the meaning of conservatism to concede the point. Just look up its definition on the Web. *Maintaining continuity with practices that develop **organically**. Preserving the **best** in society.* If only that were true! If only the best of our nature were better organized than the worst, instead of it being the other way around,[1] society would be a utopia already!

It really is as simple as that. And you know *how* we maintain organic continuity and perpetuate the best in society. ***We just stop trying to change human nature.*** We renounce the temptation to knock and knead, flog and hammer, manipulate and wrestle, wrangle and mangle our nature into something it is not. We learn to love human nature just the way it is. If that means loving it unconditionally, so be it. Unconditional love is not a luxury or indulgence. It is "an indispensable force in keeping any system stable."[2]

Only with unconditional love can we effectively mount that *resistance to change* that maintains the stability of human nature from generation to generation. Just as the happiest relationship is the one where one partner is not entangled in a frustrating struggle to change the other, the happiest race is the one where one generation is not entrapped in a futile campaign to change the other! Shades of the quagmire theory! There's nothing like politicians who get bogged down in attacking the right of an aboriginal people to govern itself according to its lights. Unless it be parents on a mission from some three-person'd God to break, blow and burn the original nature of children into something new!

Does this mean being averse to progress? On the contrary. The stability of human nature provides a highly prized *pou sto* from which a world of genuine growth

---

[1] "The problem is that the worst of our nature is better organized than the best of it." Bob Koehler

[2] Jean Liedloff, ***The Continuum Concept*** [p. 26]

and development is possible. Nature is way ahead of Archimedes in this department. Human nature is the unchanging ground from which we lever the world to a pinnacle of perfection. Far from inhibiting our freedom to grow and improve, it offers us a fulcrum from which to jiggle a cruel, inhumane political order in the direction of one that incorporates the essence of *humankind* – which is to function in a way that is as *kindly* as it is *humane*.

17. BIRTH OF *FAUX* CONSERVATISM: BURNING THE BOOK OF LOVE

Unconditional love is a buzzword in this day and age, though few grasp why it is so all-fired important. It is the *one* thing that honors our biological stability as a species. It is the *only* thing that preserves the organic continuity of our nature. It is the *one and only* thing that supports our innate resistance to change. "What interrupted our own innate resistance to change a few thousand years ago we can only guess."[1] Well no, we can do better than guess. A series of earth-shattering catastrophes led to a traumatic breakdown of our organic continuity, a disastrous rupture in our human continuum. In a despair-crazed bid to save a corrupted humanity, survivors rallied around a cult of absolute power that laid the foundation of a *faux* conservatism. And what made the shift from a genuine to a *faux* conservatism so pernicious is that introduced a theological worldview, a religious way of thinking that is so fixated on visions and flights of fantasy that it fails to "take into account the entire range of factors concerned in serving our **best** interests."[2]

This sums up the story of man. History is authored by *faux* conservatives who claim to embody the rule of the *best* but who prove themselves determined in every age to give humanity the worst of it. If this *faux* conservatism is the best at anything, it is at getting the best of the genuine article. Refusing to recognize the legitimacy of people as sovereign beings in their own right, it gets away with lording it over the human race by imprisoning its

---

[1] Jean Liedloff, *The Continuum Concept* [p. 26]
[2] Ibid.

princely heart and soul behind towering walls of repression and denial.

We don't have to be aloft in space to see the earth as one. We can be grounded in the human continuum to see the superficial differences between people melt away. I *am an incarnation of You. You are Me incarnate.* Only by being dislocated from our integral nature can we believe that I am superior to you or you inferior to me. Are these specious conceits of superiority and inferiority conducive to bringing out the best in our species? When was the last time the best in **you** came out? When you fancied yourself to be in a class better than everyone else, or when your spark of common humanity was kindled?

When a child calls out *Daddy* in a crowded mall, what father within earshot is not going to turn around, all keyed up to respond? When the cry *Mommy* rings out in the pick-up area after school, what mother pauses to wonder for whom that bell tolls – she just assumes it's for her. "There's a man who is my brother, I just don't know his name/ but I know his home and family because I know we feel the same/ And it hurts me when he's hungry and when his children cry/ I too am a father, and that little one is mine."

It's about time we all felt that way – but when is that belated chord likely to be struck? When we no longer take a dim view of any part of humanity because we are powered by the currents of our heart to turn on the lightbulb of emotional intelligence in our head? Or when we react to a traumatic break in belonging, a breakdown of bonding, by succumbing to what the philosopher of Superman dubbed the *pathos of distance* from which we look down our nose on our fellow man and reduce him to the level of Subman?

The **faux** conservatism at the basis of political rule is the most surreal breed of animal on the planet. It talks itself up as a palladium of social order, but breeds out of control in a Petri dish of rage, isolation, suspicion, hate. It bottom-feeds on conflict and division, even as its "institutions are the principal means by which conflict is produced and managed in society."[1] It runs on the principle that one man comes out ahead by leaving another behind. The conservatism of our nature thrives on the *solidarity* born of intimacy, nurture and affection. The conservatism of our institutions thrives on the *stolidity* bred of coldness, neglect and rejection. Is it likely nature endows us with splendid talents and passions to pool together for the good life for us all, so that a handful can retreat to treasure-glutted Ali Baba caves and boast of having all the things the rest of us have not?

There's no way around it. If we are to be ruled by the best and brightest in human nature, it won't be by deferring to a class of *faux* conservatives who are wrenched from the instinctual heartsprings of life. It will be by making common cause with a class of beings "who forge for themselves an *art of living in times of catastrophe* in order to be born a second time and to fight openly against the *instinct of death at work in our history.*"[2]

It's all about majoring in the art of living. The prerequisite is to take the Book of Love to heart as our highest authority. Love writes the definitive how-to book on uncovering our innermost truths and discovering they are the

---

[1] Butler D. Shaffer, **Calculated Chaos – Institutional Threats to Peace and Human Survival** [p. 6]

[2] Albert Camus, Banquet Speech at Stockholm City Hall, celebrating his 1957 Nobel Prize in literature

same for everyone. Any society that would censor these truths or suppress our right to live by them, in order to march us in lockstep with the dictates of an ersatz authority, can only prevail through a book-burning campaign. And the first book to go up in flames is the one whose values are continually being smeared as weak and degenerate, decadent and unmanly – the Book of Love.

18. THE BROTHERHOOD OF MAN
HAS ALWAYS BEEN CHILD'S PLAY

Babies may be newly born, but they don't exactly herald a novel break with the past, or signal the birth of anything really new. Babies are the oldest, most organic part of our humanity, established from time immemorial. Weighing a few pounds at birth, the human baby bears the weight of the wisdom of eons on its shoulders. Steeped in an evolutionary tradition only slightly younger than the hills, there is no whiff of the manmade about it. Rather, *it is what makes the man.* It determines our biological needs and thinking patterns, the genetic character of our feelings and the unique nature of our humanity.

When we look at a baby, what do we see? A delicate undersized specimen of humanity? Or a template of rock-ribbed conservatism, the original face of man carved in the rock of evolutionary ages. Do we see a mewling creature who cannot hold an intelligent conversation and reduces us to babbling in nonsense syllables? Or the best philosopher to teach us the meaning of life – a past master of emotional intelligence who can tell us everything we always needed to know but are too afraid to ask about creating a better world for our children? Do we see someone to be thrown into the deep end and initiated into grown-up mysteries ASAP? Or someone who knows what most of us have forgotten – how to lustily voice his genetic need to be tenderly held and take his own sweet time suckling on the milk of human kindness. And who doesn't give a crap who knows it!

In fine, when we think *baby*, do we think of someone who deserves the go-to schoolyard stigma of *crybaby* for its congenital failure to affect a stiff upper lip? Or someone whose visceral need to be enfolded in a cocoon of nurture and affection encapsulates exactly what it means to be *babied* – even as it sets the bar for that standard of warm delicious care that adults instinctively crave when they christen each other *baby*!

Society's biggest problem isn't the psychos who shoot up campuses, workplaces, movie theaters, and malls. It is the straight-arrow establishment types who, with every appearance of logic and reason, shoot down the prospect of human beings coexisting together as one big happy family. Pumping bodies full of holes is senseless and crazy, but not as senseless and crazy as shooting holes in the brotherhood of man – humanity as one vast, interconnected, cooperative body, united by the organic force of love. That doesn't qualify as breaking news on TV. *That is just accepted as the received wisdom of the day!*

And yet what is the first thing the young of our species need when they see the light? Directions to the nearest mall? A tour through the latest arms bazaar? No and no. It is to *coexist in a lovey-dovey embrace*. Nobody told them that the family of man is impossible to realize. They live to be enfolded in the bosom of family and snuggle in a touchy-feely state of togetherness. And if they happen to be denied their birthright, they can't exactly toddle off to call Congress or march on Washington. They develop a range of severe reactions from temper tantrums to personality disorders to a failure to thrive.

Here's the serendipitous part, if you could call it that. Every quality we prize in our babies exists for no other reason than to promote human bonding, closeness and togetherness! Compared to the fearless ease with which babies establish lasting bonds, adults are bumbling Nervous Nellies. The effortless freedom with which children form till-death-do-us-part commitments make parents seem like blundering babes in the woods. Who can count all the stumblingblocks grown-ups contrive to put in each other's way when it comes to forming lasting attachments. Adults have been burned so many times by a blazing coal of abuse, under the impression it is the diamond of love, they go on shying away from the diamond of love, lest it turn out to be a blazing coal of abuse. They keep each other at arm's length, petrified that if others sneak up too close, it will be impossible to hide the Scarlet Letter tattooed on their psyches: *U* for *Unlovable*. They push each other away with cruel comments and callous gestures because if they ever opened up they would be flooded by painful memories of what it was like for them to be vulnerable to cruel comments and callous gestures. And their earliest attempts at forming bonds are so consistently violated that they become past masters at violating the very bonds they go to such extravagant lengths to solemnize by walking down the aisle.

Children suffer from no such hang-ups. They are as innocent and guileless as they come. In the original state of perfection in which they are delivered to us, you don't see them wearing masks, or putting up clever facades, as they have absolutely nothing to be ashamed of. They are spontaneously open and transparent because they have nothing to hide. When my son was eight he couldn't even do a magic trick without spoiling the effect by spontaneously begging me right afterwards, "Shall I tell you how I did it?"

If babies could talk, they would echo Thomas Jefferson's words. "There is no truth existing which I fear, or would wish unknown to the whole world." They don't have their guard up at all times, living in dread of being hurt if a chink in their armor shows. With their parents there to protect them, the last thing children need is to protect their feelings. They would sooner clamp down on bladder or bowel movements than on their sad, angry emotions. Children come to us in a pure state of trustingness, with none of the standing pools of tears that breed a diseased cynicism and despair. Their sensitivity and vulnerability is the soul of their beauty. Their guilelessness and guiltlessness is the heart of their endearing charm. These are the qualities children bring to the party that enable them to bond with superlative freedom and ease. If a brain trust of sages and experts sat down to compile a laundry list of qualities that our species would need to get along as one Big Happy Family, they couldn't do any better than arrange for our young to possess the same set of precious qualities they already bring into the world in such abundance!

Did we luck out or what! What are the odds that of countless species on earth, the young of **our** species would embody to perfection the wish list of qualities needed to pull off the brotherhood of man? Billions of cellphones should ring out jubilantly with the news! The human species has struck something infinitely more precious than gold – *it has struck the Golden Age!* Twitter sites should be crashing all over! Can you believe it? Here mankind is heartsick for thousands of years for a world where we can all live as one Big Happy Family. And our conservative order of being – millions of years in the making – has already gone and laid the foundation for making our wildest dream come true!

Why, then, do we act like world brotherhood is an unachievable pipe dream, when nature evidently goes out of her way to make it *child's play*? The question isn't what kind of far-out, flighty-headed revolution is required to realize the dream of human unity. *It's what kind of crack-brained, cockamamie revolution shattered it in the first place!*

19. ENOUGH WITH THE CATASTROPHES – IT'S TIME FOR A EUCATASTROPHE

The best in human nature comes from following the line of least resistance. The worst is a hard row to hoe that goes against the grain of humanity. If you would rather not have a disruptive child, spare him the storms and stresses of a violent childhood. It is the best-kept secret of the ages. *Children who are never violated by the rod are among the best behaved.*[1] As long as the rule of the *best* is identified with *aristocracy* – an oxymoron, if ever there was one, from *aristos* = best, *kratos* = force – the lords of misrule are free to misrepresent their power-drunk orgies of slaughter and greed as the *natural* order. *Natural* and *best* go hand in hand with babies born in a tub of warm water – not with an aristocracy birthed in a blood-brimming cauldron that betrays the worst side of humanity. We are so used to thinking of formidable organizations of power and wealth as the established order, we lose sight of a basic fact. *Behind every mighty empire is an obscene history of usurpation and genocide. Behind every great fortune is a sordid dossier of crimes against humanity.* Indeed, without more of the same old kleptocratic rapine and killing rampages, it would be impossible to · maintain great concentrations of wealth and power.

If you think power represents a hardcore reality, you have another think coming. Kings and knights, palaces and castles are the stuff of fairy tales. To equate their rule with the rule of the best is to take history as a blood-soaked dossier

---

[1] Dr. Murray Straus "The best-kept secret in psychology is that children who were never spanked…."

of man's inhumanity to man and turn it into a Disneyfied fairy tale. It is on the watch of these princes and potentates, feudal lords and robber barons that the worst elements of humanity – rapacity, arrogance, brutality – come to the fore and made life ugly for much of the human race.

If life is just one damned thing after another, death would be a blessed event, not birth. But we know we are destined for more, and if we can't get it in real life, we seek it in religion and art. As lovers of fiction we are unwilling to put up with two hours of non-stop travails and tribulations without the payoff of a happy ending. How then can we allow the all-too-real story of man to be a plot of star-crossed love, dragging on interminably for thousands of years, without demanding some sort of redemptive ending?

If history is to have the fairy tale ending it deserves, we must change the ignoble beginnings in which children get their start in life. The abandoned child must bask in the eye-opening experience of having someone in his corner. The battered child must revel in the mind-opening experience of knowing that not a hair on his head will be harmed. The neglected child must rejoice in the heart-opening experience of knowing that every feeling is worthy of attention, every need worthy of being attended to. We must go back and shepherd our inner children through the valley of the shadow of death where they wander, frightened, lonely, and lost. Only then can their future be a shining city on a hill.

Eucatastrophe – a word coined in a century that witnessed the two most apocalyptic wars in history – *is a sudden turn of events for the better.* It will take a eucatastrophe to break free from a compulsion to recycle our ancient history of world catastrophes from century to century. We prove our royal birthright as sovereign beings not by pulling the sword of doom out of the firmament and dangling it over our head but by sticking it back.

But first we must have some experiential basis for knowing it is never too late to have a happy childhood. We must sign on for a remedial course in emotional intelligence as a second language and sample the paradisiac bliss that flesh is heir to. It is the prerequisite for taking the best qualities that are natural assets of the young – our prosocial genius for loving bonds and lifetime attachments – and capitalizing on them as the magic seeds for sprouting a universal network of mutual cooperation and communal accord in which each and every one of us, not a select minority, is entitled to live happily ever after.

It's just like them to pretend *living happily ever after* is a trope in fairy tales, instead of a script inscribed in the bedrock of our conservative nature. *Happily ever after* is not a byword of make-believe – it is a beatitude handed down from a peak experience of love. The terrifying end of the world cannot be the last word in religion, politics or history. In the end, everything is meant to be alright. And if it is not alright, it is not the end.[1]

Love makes our life story so enchanting we never want it to end. Love is a thing of such beauty it has to last forever. Love is the guru on the mountaintop that reveals the meaning of life. Sunsets are beautiful only if we have someone to hold us in the gathering gloom. Stars glitter like diamonds only if we have someone to spend the night with. Love is more than a memory to cherish until the end of time. It is a memory embedded in us from the dawn of

---

[1] Fernando Sabino (quoted by John Lennon and the reception clerk in *The Best Exotic Marigold Hotel*)

time. It took earth billions of years to develop a capacity to mother us – as it took us millions of years to perfect a capacity to mother ourselves. Our loving nature evolved from countless acts of nurture. Ages of loving kindness went into making us the humane creatures we are. We come into the world through a love that has been building for an eternity – and through us love goes on unfolding for an eternity to come.

When death comes to break up the love of partners or parents and children, it is a shock to our system. No matter how natural the causes of death, we spiral into anguish and grief. To mourn another's death is about all we can bear. To mourn having a hand in that death is beyond unbearable. Humans are not made for that. We think nothing of violence only because we feel nothing of love. If we opened up to the experience of love, it would be so sublime we could never ever bring ourselves to put an end to a single life.

In a funeral parlor, we are full of commiseration and condolences, sympathy and consolation. That is about our speed as humans. In the dock, taking responsibility for giving the poison, firing the bullet or closing off the wind pipe, we are choked with remorse and contrition, wracked by guilt and shame. Is it any wonder so few of us take that on? For when we do, the pain is more torturous than any punishment the state metes out.

Imagine giving a blow-by-blow report of of your murderous actions to the parents or children of your victim. Wouldn't you rather face a firing squad than the looks of horror on their faces? If God said, Let there be light, it must be the devil who said, Let there be an aching void where the light has gone out of your life. Who in his right mind wants that on his CV? Killing is the easy part compared to living with the aftermath of a beloved's death. Imagine if every soldier who killed knew he had to visit his victim's parents to console them for their loss. Or every fighter pilot who blew up a body knew he had to visit his victim's spouse or children and watch their faces crumple as he broke the bad news. Martial *courage* would turn out to be just another word for *heartlessness*.

We never want loved ones to leave us, much less depart the world. Their presence makes life so ineffably sweet that only something terrible would make us put a premature end to it. Like the end of the world. *We must fear the end of the world to put an end to life!* Isn't that the time-honored rationale for rites of human sacrifice? The love we bear one another is the veritable rock on which the temple of humanity is built. Time cannot age it as is renews itself from generation to generation. No break from tradition is more disruptive than a breakdown of love. We can survive nuclear fission but the inviolable bonds between children, parents, spouses cannot be broken without breaking our hearts.

We think of conservatism in terms of political rule. But those who rock our world are not those who control and punish us, but those who nurture and cherish us. We define conservatism as an aversion to change. But love is the only non-negotiable thing that has to be changeless in our lives. We describe conservatism in terms of an organic continuity, the rule of the best. But love is hands down the best thing that ever happens to us – and its continuity is so organic to our psychobiology that to have it abruptly end, and not go on and on, is a disaster of titanic magnitude. No loss of money, status or prestige can compare to that. *If I lose my fame and fortune, and I'm homeless on the street, and I'm sleeping in Grand Central Station, it's okay if you are sleeping with me.*[1] Strip away everything else and the conservation of love is the only thing that keeps us going. If love could be distilled into three little words, they would be *Happily Ever After.* The child as we know spells love as T-I-M-E. The child within us spells love as T-I-M-E-L-E-S-S.

---

[1] Jerry Duplessis and Wyclef Jean, "My Love is Your Love", Whitney Houston

III. THE TRASHCAN BABY – A REVOLUTION IN CHILD-RAISING

"When I became a radical nearly 70 years ago, we ran 'the risk of seeming ridiculous,' as Che Guevara put it, if we thought Love had anything to do with Revolution. Being revolutionary meant being tough as nails, committed to agitating and mobilizing angry and oppressed masses to overthrow the government and seize state power by any means necessary."

— Grace Lee Boggs, **The Next American Revolution** [2011]

Where love rules, there is no will to power; and where power predominates, there love is lacking. The one is the shadow of the other.

— Carl Jung, **The Psychology of the Unconscious**

20. THE BLOOD OF MARTYRED EMPERORS SEEDS THE CHURCH OF IMPERIALISM

The setting is a fortified mansion in a city near the frontier of Europe and Asia. The sound of shelling signals an approaching army. The family inside, awakened at midnight, is told to pack for a journey. They are herded into a basement – mother, father, four young girls, an infant son – along with some family retainers. A dozen men rush in with pistols, in various stages of drunkenness. With a frightful din, an orgy of shooting ensues. The mayhem lasts twenty agonizing minutes. When the smoke clears and shrieks subside, the parents seem to have been killed instantly. The children suffered before dying. The baby, moaning and writhing over his dead parents' bodies, was dispatched with a revolver. The wounded girls, watching their parents die, were stabbed with bayonets. Bits of flesh and blood clung to the walls. Streams of blood pooled on the floor.

The massacre sent shockwaves through Europe. For this was no ordinary family. It was Russia's royal family – the last of the Romanov dynasty. Nor was it an innocent family. It was Tsar Nicholas II, the Tsaritsa and their heirs – representatives of three centuries of autocratic rule. What appears at first blush as a ghastly mass murder is the ringing blow of a *revolution* waged against a hated tyranny. Nicholas was not exactly a benign ruler, and his people were desperate for change in 1918. God may work in mysterious ways, but the ways of his royal proxies on earth are nothing if not despotic, rapacious, cruel, and capricious beyond measure.

The 19th c. is bookended with the buzzword *revolution*.

In the preceding quarter century, there are two back-to-back revolutions – American and French. In the succeeding quarter century, there is a third – the Russian. Freedom becomes the rallying cry, monarchy becomes a dirty word. The target is the crowned heads of autocracy. If not actual kings, then supporters of monarchy, monarchists or royalists. If not emperors, then supporters of imperialism. There is something screwy about this. One could call it the law of unforeseen consequences, except that it happens so consistently one can see it coming from miles away. Lethal attacks on monarchs are not so much revolutions as *revolts* that lead to the installation of another sort of monarchy.

Was the French Revolution against a reign of aristocratic excess? Tell that to billionaire owners of luxury villas on the French Rivera. Did the boot heel lift off the neck of the common man after all those aristocratic heads rolled? Ask the military general who had the arrogance to crown himself emperor after a spree of military conquests left him master of continental Europe.

Or take the Russian Revolution. When peasants and soldiers revolted against the emperor, the Bolsheviks established a dictatorship, in many ways more efficient and absolute than the Tsar's. And as the Soviet Union tightened its grip over far-ranging territories and multiethnic peoples, it became a textbook empire, feared, rightly or not, as a specter of world domination.

And let's not get started on the American Revolution. A nation ushered into being amid ringing tributes to freedom and independence is founded on encampments of slave labor and genocidal wars of expansion against native populations. No sooner is the continent in its blood-soaked hands than it set its bloodthirsty sights overseas. It starts with a few colonies in the Caribbean and Western Pacific before WWI, and by the time WWII wraps up, it goes on country-shopping spree to become the mightiest nation on earth, installing mad-George dictators here and there and planting a thousand and one military bases everywhere. Look where its revolutionary war against the British empire got it barely two centuries later – uncontested heir of the empire on which the sun never sets. *We're an Empire now, and when we act, we create our own reality.*[1]

One can't help thinking there is something *phony* about such Revolutions. They make a great show of attacking the figureheads of tyranny and imperialism. They blow them away with muskets and cannons, chop their heads off with guillotines, mow them down with firing squads. When the smoke clears the tyrant king is dead. The system of tyranny has a new lease on life to grow from strength to strength! Corpses of the imperial family are doused with sulphuric acid, burned and buried, but imperialism rises like a phoenix to bestraddle the earth as never before!

What gives? How did a country born in a paroxysm of anti-monarchist militancy become *de facto* monarch of all it surveys? How did the assassins of the imperial family become the architects of an imperialist regime that spread its tentacles over half the earth? It's fair to say that the forcible overthrow of an existing political order falls into the category of a *faux revolution* – that is, the overthrow of an oppressive order that makes way for one that is equally oppressive, if not more so. At the same time, the target of faux revolution is not even a *genuine* conservative order. It is a *fake* conservative order where a tiny minority lords it over

---

[1] A high official at the court of George W. Bush, acknowledged to be Karl Rove.

a vast majority. Indeed, all that this historic pattern of *faux revolutions* succeeds in doing is giving a *faux conservatism* a brand new, extended lease on life! Sounds crazy, doesn't it? Instead of driving a stake through the heart of faux conservatism, the faux revolution makes a mockery of its noble principles by ramping up the faux conservative order and perpetuating its oppressive rule with a vengeance!

Did the end of the Austro-Hungarian empire put an end to imperialist fantasies – or spawn a dynasty of *imperialists of money* who subvert democracy by taking the law into its hands? Did the American Revolution disabuse us of imperialist conceits like the Monroe Doctrine? Or did not the fall of Britain's Empire pass the imperialist torch to its heir apparent, the United States?

The vendetta against royalists is like the persecution of early Christians – the more we mow them down with gunfire, or chop their heads off with guillotines, the more the worship of power grows by leaps and bounds and the cult of world domination mushrooms in scope and sophistication! *The blood of martyred emperors seeds the church of imperialism.* From the ruins of feudalism spring the lords of finance to impose debt serfdom. America hasn't had a king for a quarter of a millennium, yet Americans continue to be vassals of the kingpins of corporatocracy.

21. THE ONLY THING WORSE
THAN BEING EATEN BY A MONSTER
IS IDENTIFYING WITH IT

For an institution like monarchy, billed as the organizational center of earthly/heavenly order, an awful lot of monarchs have lost their lives or thrones. And not just in recent times. The roll of deposed, exiled or executed monarchs stretches into the recesses of the past.[1] If kings and their latest avatars, kingpins of banking and industry, are such mainstays of world order, why the eternal groundswell of raging discontent over their rule? Even in 2100 BC, when kings passed for gods, the king of Akkad spent most of his 40-year reign quelling revolts. *The fact is not all of us are royals, but all of us expect to be treated like royalty.* And when we are not, we cannot help but be swept up in a state of high dudgeon, and no power on earth can keep a lid on it.

The real story of civilization is not the downfall of hundreds of royal pretenders who had no business claiming a fictitious kinship with divine authority. It is the dethroning of billions of our kind who are *born* to be treated like royalty. It is more than a crying shame the entire human race falls into a class of noble émigrés, exiled from their natural seat of glory by gory revolutions in childhood. It is the crime of all crimes. The ghosts of murdered kings and queens stalk the corridors of history because we have never faced up to it. Democracy still struggles to rise above a charade because we fail to redress it. It's not a matter of removing the deluded souls who claim to be of royal birth

---

[1] Check out Wikipedia's lists of monarchs who have been dethroned or assassinated in the past.

but of failing to guarantee the young of our species their royal birthright.

Royalty is not a bad thing per se. The reverence and devotion that is the currency of royal treatment is the high-water mark of love. *Where it is directed makes all the difference.* Adoration lavished on persons with delusions of grandeur goes to their **heads**. There it fuels the megalomania that leads them to separate themselves from mankind and act as though others are of no account. Adoration lavished on the small and vulnerable goes to their **hearts**. There it empowers them to stay connected to their kind and tap into the fellow feeling that takes others into account.

There is no need to anoint a man's head and bow to him as king. There is a need to wash a baby's bottom and ooh and aah over him. It's not natural to exalt one person above others, but nature makes it our sacred duty to raise a child as a Visiting Dignitary. It's not rational to hold up one person as a pillar of order and life, treat his wishes as commands and smother him with luxuries. But we have every reason to give children a hero's welcome and hold them in high regard as sources of beauty and life and pamper them until they feel at ease in their skins and at home in the world. It is the height of rationality to fulfill their dependency needs and fill their days with laughter and song until they come of age as sovereign beings in their own right.

The Child as Guest of Honor lays the foundation for the world as a hospitable place, providentially entrusted with making our stay on earth a pleasant and fulfilling one. We don't have to rattle around in palaces. We just have to find a place in another's heart. Nobody has to jump to our commands. They just have to smile as we jump up and down in delight. We don't have to make anyone's life a misery. We just have to be around those who take joy in our lives. We don't have to stand over people and look down on them. We just have to have people stand up for us and look out for us.

Before a crown was an ornamental circlet of gold to be worn on a person's head, it was the top part of a baby's head pushing its way through the vulvar ring. Infants don't have to beat their heads on a wall to be raised to occupy the throne of thrones. All their heads have to do is to **crown** in the final stages of labor for the occasion to be heralded by a joyous coronation that invests them with their birthright of royal treatment. They don't have to sit on a peacock throne to get homage and devotion. They just have to sit on our laps as playful objects of affection and attention, protection and respect, approval and understanding. Not gold and silver, ivory and jade, pearls and rubies, but tender caresses, doting glances, warmhearted smiles, and indulgent nods, are the raw materials for assembling the throne of love. This throne is the foundation for the primal order of our being, which for a lack of a better word we can call *archeo-conservative*.

Long before a historic pattern was laid down of overthrowing the pseudo-conservative order with paroxysms of human violence, our *archeo-conservative* order was overthrown by paroxysms of celestial violence. That's the funny thing. If celestial violence didn't forge a royal partnership between humans acting as war-crazed kings and gods acting like wrath-crazed parents, revolutionary forms of social violence wouldn't be necessary to unseat these quasi-divine rulers and replace them with good parent types, amenable to human concerns and considerations.

The doomsday curse is to become what we hate and fear. The impetus behind *faux* revolutions is a cry from the heart, but the heartless violence they engender reinforces the status quo. The response to child abuse is an eruption of fury, but the revolution it sets off is another cycle of abuse. When we can no longer marshal the best of our nature to mount a meaningful resistance to monsters, we become conduits through which the worst of our nature uninterruptedly flows. The most senseless violence comes from those who are shocked and awed into insensibility. The most vicious outbursts come from those who are too demoralized to stand up to abuse. The most virulent attacks of machismo come from those who are too neutered to fight for their humanity.

Our archeo-conservative nature is sustained with love. As assaults on our royal person goad us into taking up arms against our tormentors, we beat a retreat from the ground of our humanity, and are lulled into sleeping with the enemy. There is hope as long as humans are moved to reclaim lost parts of the self with the patriotic fervor of revanchists, determined to regain lost parts of their country. There is little or no hope if they are willing to surrender their innocence, integrity, worth and dignity in order to deify the forces of chaos and destruction, bow down and worship them as gods, and treat them as transcendent authority figures, to be obeyed at all costs. Revolutions start out as a revolt against the bad king, but generally wind up embracing the bad king to make common cause with him. For the worst of it is not to be devoured by a monster and digested in its maw. It is to be assimilated by a monster to the point of identifying with it. The archeo-conservative order that is overthrown need not be that of the little baby. It could be that of the grown man – as in the case of Job, where the blueprint for such a revolution is to be found.

Job starts out with the expectation that God as sovereign ruler of the universe is the Good Parent incarnate – a fountainhead of justice and compassion, a providential bastion of order and sanity. By the time he is done being bedeviled by the forces that blast his home, bereave him of family and obliterate the pillars of his existential security, Job winds up making common cause with a wrath-crazed tyrant who terrorizes him out of his wits. This is the gist of his story. After a long period of agonizing over what to make of this preternatural abuser of a God, Job makes nice with him. He is the most tragic example in the annals of existence of that hoary dictum: *if you can't beat them, join them.* The best of Job wants to mobilize justice and reason to revolt against the dictates of a sadistic monster. But with no one to support him, play his advocate, or help his case, Job in his desolation gets the worst of it – and like the protagonist in **1984**, facing his worst nightmare, he breaks down to tearily embrace his tormentor in an orgy of self-loathing.

So what if this God is a demonic predator of cosmic proportions, gnashing his teeth and breathing fire. So what if this God rides a crooked serpent, a celestial dragon that bears an uncanny resemblance to prehistoric descriptions of a world-destroying comet. Job drops his resistance like a hot coal and identifies with a God who is himself identified with the forces of chaos and destruction. His innocence and goodness, his very conscience is cauterized out of him like an open wound. And the only path of survival left open is for Job to abdicate his human nature, as far as he can, and go over to the Dark Side

"We convert him, we capture his inner mind, we reshape him…we bring him over to our side, not in appearance, but genuinely, heart and soul."[1]

In the end, Job takes an abysmally dim view of the very things that make man magnificent, thereby providing the optics for clarifying our reading of history. The monstrous apparition of a God to whom Job swears allegiance is a model of the king that rules society. *Everyman like Job is henceforth ruled by the archetype of a god monster that is not bound by a calculus of human reason and decency.* At first it is couched in the mythic terms of a divine force, or **kratos**, incarnate in the warrior-hero, the supernatural beast of prey, the malevolent sword of a comet.[2] And how apt **kratos** was to prove as a suffix for government (**kratia**, **cracy**) is evident thousands of years later, when an iconic work of political philosophy formally recognized the amoral power structure of the state by bestowing on it a shocking title drawn from the Book of Job: **Leviathan**!

---

[1] George Orwell, **1984** [p. 210]
[2] Rene Girard, **Violence And the Sacred** [p. 264]

22. THE REBRANDING OF MORAL ANARCHY AS 'CONSERVATISM'

Nothing in human nature supports a monopoly of kings. That is why you have to go beyond a human frame of reference to validate the politics of kingship. Kings cannot carry off their delusions of grandeur without affecting a kinship to gods. This wouldn't be so bad if gods embodied the best in humanity. But in fact the gods whom kings modeled themselves after were planet-destroying monsters. And humans could not very well emulate gods who initiated a catastrophic breakdown of *cosmic* order without suffering a disastrous breakdown of *mental* order.

Doomsday gods are nihilistic in essence. They are wrath-crazed monsters of celestial origin, presiding over a nihilistic dissolution of the limits and boundaries that define the **terrestrial** condition. By extension, their royal representatives on earth were wrath-crazed monsters who personified the nihilistic dissolution of the limits and boundaries of the **human** condition.

Is it a coincidence that paleontologists dubbed the most fearsome specimen from the Age of Dinosaurs *Tyrannosaurus Rex?* It's almost as if men could conceive of nothing more monstrous than a *tyrant king*! We know an errant comet or asteroid wiped out the dinosaurs millions of years ago. Who would have thought the same sort of celestial catastrophe that killed off T-Rex in prehistoric times, gave birth to the monster king that terrorized the human race in historic times!

Kings cannot bear an affinity to gods without cutting loose from the limits of humanity. Man is by nature a social animal, as Aristotle pointed out long ago, and he cannot extravagate beyond the human pale without being denatured into a god and/or a beast. Isn't that why wrath-crazed monsters still roam the earth? The same type of cosmic disasters that marked the end of the road for T Rex forced man to take a wrong evolutionary turn and wind up in a world where the descendants of T Rex in human form still rear their ugly heads and breathe down our necks.

The ancient monarchies of Africa were the first to vividly dramatize this Jurassic model of kingship. We think of a king as a superhuman pillar of moral order and law. In fact, nothing can be put past a being who refuses to abide by the limits of humanity. The king is traditionally free to overstep moral boundaries and be guilty of unspeakable actions that highlight his affinity to a *savage beast*. On occasion, he is even "required to commit all the forbidden acts that are imaginable and possible for him to commit."[1] *Trespass* or *transgression* implies limits beyond which it is taboo to go. If there are no human limits for the king *qua* god, he is in a position to trespass or transgress with impunity, and in point of fact, is expected to embody the "paragon of transgressors, the man who holds nothing sacred and fearlessly assumes every form of hubris."

The king's resemblance to a savage beast, an amoral monster would be an aberration confined to myth and ritual – were it not documented in nauseating detail by the historical record. Even a cursory survey of kings and emperors shows there is not a single crime they are not guilty of in the course of their reigns. The notorious depravities of pagan Roman emperors reads like a veritable encyclopedia of crimes and transgressions. And as the natives of the Americas can attest,

---

[1] Rene Girard, **Violence and the Sacred**. Quotes in this paragraph are from [p. 110] and [p. 105]

who needs a Nero and Caligula when you have a Columbus and Cortez who came as royal emissaries in the name of the Roman Catholic kings of Spain, but who may as well have come as human avatars of the *comet* in terms of the pestilence and war they unleashed on the New World.

Thus the *sacred* tradition of kings as paragons of transgression meshes perfectly with the *secular* tradition of kings as bywords of lawless voracity and debauchery. *History* as an encyclopedia of crimes against humanity is the *mythology* of sacred monarchy in action. Why should we be shocked by this? After all, if history is little more than a register of the crimes, follies and misfortunes of mankind, who is responsible if not the kings and emperors who actually preside over history and leave their bloody mark on it? Who is to blame if not the so-called *dei majores* who arrogate the right to run mankind's affairs? The surprising thing is not that kings who make a state religion of hubris should show such flagrant regard for the bounds of decency, common sense and reason. It is that society should cower in awe and dread and abject humility before these royal incarnations of *Leviathan. He beholds all high things: he is a a king over all the children of pride.*[1] Man actually looks to these monsters to bring order and stability to society when they clearly embody a sociopathic contempt for the integral laws and limits of humanity! Man's Job-like plight down the ages resembles nothing so much as that of the little girl in the grip of a terrifying nightmare, who cries out for her mother to come and save her from the clutches of the monster

---

[1] Job 41: 34

– only to awaken and discover that her mother herself is the monster!

A conservative label denotes a tradition of continuity and convention. How can it apply to royal disrupters like kings who rupture the norms of the human continuum by fancying they are gods incarnate? How can wannabe gods whose lifestyles are a monument to extravagance and excess serve as bastions of conservatism? How is it possible to launch as public guardians those whose delusions of godhead are a symptom of mental illness, and parlay their mental instability into a guarantee of political stability! To market those who make a career of going off the deep end as captains who can be trusted to steer the ship of state – is that the most mind-boggling public relations campaign in history or what! Why are we not blown away by this feat of rebranding? The act of reading history is not an exercise in objectivity – *it is the most wildly successful experiment of all time in thought control.* How else can we follow the criminally insane actions of a class of beings that exhibits a bizarre break from reality, a disintegration into moral anarchy and every sort of material debauchery, and be manipulated into thinking of them with reverence and respect as *pillars of law and order* and *palladia against anarchy*!

Edward Bernays is hailed as the father of public relations. His mission was to orchestrate the case for war, to seduce ordinary people into participating in an orgy of mass murder and mayhem. And he pulled it off by taking a hardworking, law-abiding, high-achieving people like the Germans and casting them as barbaric Huns, a scourge of humanity that deserves to be despised and destroyed.

As sensationally as he killed it, Bernays has nothing on the author of the *Book of Job*, who has the distinction of writing the groundbreaking bible on the art of propaganda. He takes a draconic apparition of a comet that aggressively attacked the earth, in violation of every cannon of human justice and reason, and achieves the remarkable feat of turning this monster into the type of god king who is to be blindly worshipped and obeyed. And in the process, he takes an Everyman of a Job who is a paragon of virtue, as innocent and upright as they come, and vilifies him so completely that he turns him into the prototype of the *canaille* – a cowering cur who comes crawling back to lick the hand of the master who makes a point of repeatedly kicking him. *Behold, I am vile; what shall I answer thee? I will lay my hand upon my mouth.*[1]

---

[1] Job 40: 4

23. THE DIVINE RIGHT OF KONGS

By force of habit we speak of an American *Revolution.* It wasn't exactly a radical break with the past, because the order of kings it was fighting was anything but conservative. It was a monstrosity that flouted common sense, defied the bounds of common humanity and wreaked havoc on society. As much as the Founders failed to surmount the prejudices of their class and era, this much can be said for them. *They recognized monarchy as the 100 foot, 1000 pound gorilla in the room.* Kings had presented a problem throughout history – and as long as they ran amok, there would be no peace or prosperity for the people. So the Founders saw it as their task to repurpose government as an enclosure to cage up the savage beast. To the rest of the world they were attacking the sacred seat of civilized power. In hindsight, we can clearly see what they perceived dimly. *They were emissaries from an enlightened New World, bringing civilization to the island jungle of an Old World that was ruled by a monster that the natives worshipped as god.*

How exactly did they carry out their mission to capture and imprison this potential destroyer of men and cities, revered as the guardian of the realm? By compartmentalizing government into a system of checks and balances. The offices of president, judge and legislator were set up to check and counterbalance each other, ensuring that no branch of government got too big for its britches. As proud as later generations are of their work, they badly miscalculated. They underestimated the ferocity of the Monster they were up against. It had deep-seated historic roots in the ancient cult of **kratos** – the predatory principle of force and domination identified with divinity.[1] The mere idea of KING doesn't begin to cover it. To take the full measure of the Monster you have to add the word KONG. Only then can you get an inkling of the killer gorilla in the room. The constitutional grid that the Founders built to confine this monster was less an iron cage than a cheap plastic one. It was just a matter of time before the beast bust out to wreak havoc on the human scale of the world around it, dooming the experiment in democracy to fail.

In ancient times, the currency of power is divinity – the mythic stuff that is the basis of political power. In modern times, it is money – the abstract stuff that is the basis of economic power. A constitutional system of checks and balances sought to limit the traditional exercise of power by repealing the divine right of kings. But even as monarchs are downgraded to mere mortals, money is exalted from a symbolic unit of exchange into a providential entity from which all material goods and blessings flow. Though the right to issue and print money belongs to Congress, it is arrogated by a banking cartel that parlays the divine right of capital to become *de facto* kings of society. The Saint Georges of the American Revolution fought a valiant battle to vanquish the Money Monster and chain it up in service to the people, but it proved too rabid to contain. It tore loose from its constraints, reared up on its hindquarters and went on rip-roaring rampage from which the world never recovered. Bankers found a way to reduce the politicians' system of checks and balances to a house of cards. *Give me the power to print money and I care not who makes the laws.* And democracy was never the same again as subsequent generations of Jobs

---

[1] Rene Girard, **Violence And The Sacred** [pp. 263-264]

were confounded by the latest incarnation of Leviathan: *The Creature From Jekyll Island*.[1]

The cult of **Kratos** or divine force groomed kings as King Kongs – oversize forces for channeling the wrath of gods, consuming people through blood sacrifice and confiscating their common wealth through taxation. The kingpins who are their heirs are not all that different. As long as their official brand of *Kratos* prevails, *times are always going to be tough – in addition* to being *hard, cruel, violent*[2] – for they use the unlimited power of money at their fingertips to suck the lifeblood of the economy, sack the wealth of nations, and drain the health and well-being of peoples. Their monopoly of money gives a whole new meaning to the old ritual of sacrificing human flesh and blood to phantom gods. Instead of money being the servant of man, mankind winds up sacrificing love, morality, peace of mind and life itself for the sake of a paper symbol. As kings are synonymous with warriors, bankers are synonymous with warmongers. Wars then as today are bloodthirsty mechanisms for aggrandizing territory, amassing resources, and occupying the world. Heads of state are mere masks for the **death's heads** that actually rule mankind through poverty, disease, bloodbaths, bankruptcies and depressions. And if there is any homeland they can be said to be true patriots of, it is their natural home and habitat on **Skull Island**.

Checks and balances that may have curbed the power

---

[1] The title of a seminal exposé on the Federal Reserve System by G. Edward Griffin (1994)

[2] Rene Girard *Violence And The Sacred* [p. 263]

of kings as representatives of the Almighty, abjectly failed to curb the power of bankers as representatives of the Almighty Dollar. While it is compulsory for the rest of us to balance our checkbooks, those in control of the banking system make a mockery of the whole system of checks and balances. Gods of capital spurn regulatory checks and override regulatory balances. Refusing to be policed by rules of common decency, they run riot in a gladiatorial arena known as the free market. Their untrammeled pursuit of greed puts everyone else's life and liberty at risk and pursuit of happiness to shame. Their fetishization of profit, unchecked by any regard for family values, be it of the individual family, family of man, or family of mother earth, treats freedom and democracy as the gum to be scraped off the imperialistic bootheel of capitalism. Money masters who pay the piper circumvent the intent of the framers by getting the executive, legislative and judicial branches to march in lockstep to their tune. A constitutional system of checks and balances is no match for a corporate system of interlocking directorates. And at least one Supreme Court justice had the grace to own up to it.[1] The runaway concentration of wealth is a spreading cancer in the organs of democracy.

Horror movie iterations of King Kong never end happily. Clambering atop the pinnacle of temples of commerce, the giant ape is brought ignominiously crashing to earth. In real life, the money masters bestride the landmarks of military and financial might as if they own them, which

---

[1] "We can either have democracy in this country or we can have great wealth concentrated in the hands of a few, but we cannot have both." Attributed to Louis D. Brandeis

they do, and soar from triumph to triumph. If anything is brought crashing down, it is the skyscrapers themselves, in a spectacular piece of political theater that advances their agenda of marital law and world mastery. *Bring down a building or two, prop up an empire.*[1]

The history of kings is an open book when it comes to all the mischief and misery they spread. A bare-it-all historical accounting has yet to be made of the Kongs of Capital whose rampages have destabilized continents, crashed entire economies and ignited world conflagrations.

---

[1] Without 9/11, there would be no Orwellian war waged on the American people – no Patriot Act, no Homeland Security, no explosion of the surveillance state. There would be no open-ended war on terror waged on peoples of the Middle East and Africa – no wars of aggression and decades-long occupations of countries like Afghanistan and Iraq, no U.S.-led invasions, coups, assassinations of leaders in countries like Libya, Pakistan, Somalia, Syria and Yemen.

24. OCCAM'S RAZOR – SUPERWEAPON OF ORGANIC CONSERVATISM

When British politician Edmund Burke penned his valentine to the 'conservative' order after the fall of the Bastille, he declared his undying love in these terms. "We fear God; we look up with awe to kings; with duty to magistrates; with reverence to priests; and with respect to nobility."[1] He can be forgiven for holding these sentiments as natural and true, and reacting with alarm and trepidation to attempts to subvert them. But then he would have to be faulted for glossing over the delusions of godhead that serve as the basis of kingship and falling into the absurdity of belying his own conclusions. "When men play God, Burke said splendidly, presently they behave like devils. The guillotine and other horrors of the revolution bore him out."[2] Someone should shake the man by the shoulders and shout in his ear. *No one played god longer and harder than kings! And no one bedeviled society with a more infernal record of mischief.*

The new-fangled conceits of the French revolution, Burke laments, run contrary to human nature. *That ship sailed a long time ago.* The whole dispensation of culture heroes and kings originated in a gaping wound in the human order. The rules imposed by sacred myths under the tutelage of Old Masters necessarily override the givens of our existence. Burke argues that revolutionaries gravely

---

[1] Edmund Burke, **Reflections on the Revolution in France** [p. 223] Harvard Classics

[2] Edmund Burke, **Reflections on the Revolution in France.** New York: Arlington House. pp. xix.

err in spurning a repository of wisdom built up over aeons of experience. *Because a person's moral estimation is limited, people are better off drawing from the general bank and capital of nations and of ages than by trading on their own stock of intellect and understanding.* Yeah, tell that to the priests and kings who trumpeted their superhuman power and authority over the laws of human nature, painstakingly accumulated over millions of years of evolution!

The radical break from the conservatism at the heart of man provides the *opening on the transcendent*[1] through which kings, shamans, and magistrates are birthed as guardians of society. What legitimate need is there for this breed of wannabe gods? The last time we checked, humans are quite capable of living in community within the duly constituted limits of their nature. There is no inborn drive to flout our species-specific limits. The conservative order of those who steer society as gods incarnate is a misguided attempt to reinvent the wheel. Given a perfectly good system for keeping social order, the focus should be not on deterring kings or criminals, but on what drives people beside themselves to play kings or criminals. Beheading kings or herding them into constitutional pens doesn't cure an obsession to be king, any more than lethally injecting criminals or locking them in correctional cells cures a compulsion to be a criminal. When we understand what lies behind the impulse to violate human bounds, we can get in front of it. Only then can we put behind us the imperative for pseudo-guardians to save us from pseudo-monsters.

We start with a logical tenet called Occam's Razor. "Entities shouldn't be multiplied unnecessarily." We all

---

[1] Mircea Eliade, **The Sacred And The Profane** [pp. 26, 180, 211-212]

know colloquialisms to this effect. *Keep it simple. Cut to the chase. Get down to brass tacks. Quit beating around the bush.* In other words, let's shave off the frivolous, superficial layers of explanation, the distracting, digressive steps that stand in the way of a clear, concise, basic understanding of a problem. Burke's passion for order made him sympathetic to the American Revolution. Its constitution clicked with his principled opposition to the *kratos* or force that gathers all power to itself. But the whole business screams for a salutary application of Occam's Razor. Why go to the trouble of devising a polity of checks and balances when humans come with an organic system of checks and balances for sharing their planetary home? Having the other there for us is the only existential check we need to happily coexist. Being enfolded in the bosom of community is all it takes to keep us in balance. Prosocial bonds and filial attachments are stronger than chains of iron. We don't need kings who suffer a psychotic break from reality to ground us in our humanity. Nor do we need a doctrinaire writ cobbled by a committee of men of wealth and learning to organize the body politic. To retool government by the numbers according to a Cartesian charter is to dance around the basics. Our living system is the only real government we need to be the best we can be. Our psychobiology – not some artificial system reached by debate and deliberation – is the prime guarantor of liberty, equality and fraternity.

If the human organism has its own set of biological checks and psychological balances, the experience of love is all it takes to live by them. Technology builds us houses with four walls and a roof. Love empowers us to be at home with ourselves. As others are there for us, we learn to be present to ourselves. As others are sensitive our needs, we are sensitized to theirs. As others treat us kindly, affectionately, respectfully, we grow to treat others with kindness, affection, respect. Maturity is a matter of give and take. The family is our crowdfunding source, investing the seed money of love in us before we turn a profit, as it were, and develop a green thumb to make love flower and fructify all around us. By treating us as Guests of Honor, love shows us how to treat others as Guests of Honor. The law of reciprocity is key to the Golden Rule. Coming into the world naked, needy, helpless, vulnerable, we are sheltered as precious beings and nursed on the milk of human kindness. That is how we grow up to do the same for members of the family of man, no matter how naked, helpless, vulnerable or needy *they* may be. We don't need threats of hellfire to love one another. We just need heaven to lie about us in our infancy.

Ah, but there's the rub. For heaven to lie about us in our infancy, the heavens must stand firm over our heads! If the heavens tumble down in fiery chunks, how can they serve as a blissful haven for our young? *Caring* and *protecting* are as central to our relationship to our young as they are to our relationship to the environment. *Conservation* is the official care and protection of the earth, as parenting is the care and protection of the soul. Our mothering of *psychological* resources depends on our husbanding of *natural* resources. When marauding comets, errant asteroids, supernova explosions, what have you, ripped the terrestrial fabric of the universe, they set fire to forests, vaporized oceans, gutted the earth. Doomsday was an all-out assault on *conservation*. Malignant forces in heaven were no longer invested in caring for the earth and protecting its fragile beauty. And as most of mankind is wiped out, a remnant is so wracked with trauma that it is no longer in a position to care for and protect the fragile beauty of its young.

It's a simple case of cause and effect. The apocalyptic breakdown of *conservationism* for mother earth had a disastrous effect on the *conservatism* of her human progeny. Human nature is safeguarded no longer by life-giving forces that cherish and protect, but by doom-dispensing forces that judge and overpower. It is a reversal of astronomical enormity! Whereas the love that conserves human nature is innate to it, it is now believed the only way to maintain the natural order is by channeling a divine wrath that erupts from above and beyond it. The *love* that is integral to human beings is now superseded by the *wrath* imputed to a class of imaginary beings. The life-giving order bound up with caring for and protecting our young is supplanted by a death-dealing order bound up with the homicidal violence and genocidal aggression of gods.[1]

Parents channel the reproductive love of humanity, kings the destructive wrath of divinity. Parents uphold social order through an innate love of infants, kings through an internalized wrath of the gods. Kings wrest royal treatment through force of arms, infants receive it in loving arms. A baby cannot mobilize a world of resources for its needs, yet no king has a greater superabundance of love. A king borne aloft on the shoulders of others is conditioned to overawe and tyrannize. A baby raised up by love to where he belongs is under no compulsion to lord it over others. That is why the *philosopher king* is an utter oxymoron. The love of power embodied in kings rules by reducing people

---

[1] Rene Girard, **Violence And the Sacred** [p. 263]

to helpless children. The love of wisdom embodied in philosophers is exemplified by the wisdom of love which rules by raising children to be mature adults.

Kings are institutional counterparts of parents – but with a schizoid twist. Instead of harnessing mechanisms natural to our species to maintain a communal solidarity, they use *preternatural* forces of death and destruction to stand apart in a class by themselves. Instead of cultivating prosocial bonds with unconditional love, they invoke a doomsday paradigm to bind society with penal judgments and punitive laws. Kings channel a war of the worlds through political institutions to become the foremost purveyors of war as a means of enforcing divine claims and royal prerogatives. The warrior-king is a tautology. *War is the natural state of kings.* Martial violence and aggression are the stock in trade of royalty. And the fetish for militarism comes not from kings serving as oracular guardians of mental and political stability, but from incorporating the random violence and destruction of *cosmic* nature to function as archetypal loose cannons.

The *wrath of God* is anthropomorphic shorthand for a cosmos gone mad. How can a cosmic pandemonium created by gods preternaturally *mad* at man be conducive to a sane society? How can a meltdown of nature do justice to a species that evolved solid prosocial tools and bonding mechanisms for fostering solidarity? What's the sense of deploying an arsenal of apocalyptic wrath and violence inspired by Judgment Day, when we can bank on a repository of common sense and decency built up by an evolutionary legacy of unconditional love? Must a social order be beholden to faux-guardians to follow the visionary scriptures of a theological tradition, when honest-to-goodness parents are obliged to follow the visceral script of a biological heritage? As a tool of revolution wrath has never been successful at liberating us from the scourge of kings and gods. Only love can do that by raising us to be sovereign beings in our own right.

Just think of how traditional ideas of kingship and war make a blunt mockery of Occam's razor. Kings have to pretend to be gods to teach us lessons with fires of wrath, when parents can guide us with the light of love simply by being themselves! Commanders-in-chief have to flaunt apocalyptic spears to terrorize man into line, when parenting techniques are ready to hand to organize communities along prosocial lines. Must comets blazing a path of destruction across the sky morph into bolts and cannonballs, and bolides from trebuchets morph into missiles from nuclear silos? With all the advances made since the Bronze Age, here is the Royal Air Force in the middle of the 20th century, still faithfully reenacting the biblical holocaust of Sodom in bombing campaigns like Operation Gomorrah, raining thousands of pounds of incendiaries on cities like Hamburg and Dresden to create infernal firestorms that suck tens of thousands of souls into hell!

Remember that the next time institutions of power or force are spoken of as *conservative*. *Conservative* is not parsimony of *expenditure* – it is a parsimony of *explanation*. It is not cutting budgets – it is cutting away the twisted, torturous rationalizations that keep our lives from being simple and sweet. Occam's razor is the only weapon genuine conservatives need to slash over-blown, over-the-top forms of conflict resolution, and get down to the skin-to-skin contacts, face-to-face interactions and heart-to-heart communications that foster good will and peace on earth.

What is conservative about settling our differences

with doomsday weapons? This is not conservatism but the death throes of conservatism. It is an operatic flight into cloud-cuckoo land. It is Götterdämmerung elevated to statecraft. It is the ridiculous conceit that our prosocial capacity, our stockpiles of EQ are all used up, and there is nothing left but to chant propaganda slogans and whip ourselves into orgiastic dances with cosmic wolves to get each other to be nice and do the right thing! It implies that man has so far lost track of tried and true remedies for keeping his house in order that he must invite an Old Testament God to come in and tidy up the living room![1]

---

[1] Dwight MacDonald, "The 'Decline' to Barbarism". *The Atom Bomb, The Great Decision* ed. Paul Baker [p. 86]

25. A REGICIDAL CAMPAIGN
SPANNING THE AGES

We could cry crocodile tears for the overthrow of a *faux* conservatism that has been around for thousands of years – or cry real tears for the overthrow of a genuine conservatism that has been around for millions. *Faux conservatism* is reinforced by the *faux revolution* it triggers. As for the equable, equitable paradise that is the outgrowth of *genuine* conservatism, there is no *earthly* reason to revolt against it. But in due course there is a *celestial* one. By dint of countless revolutions around the sun, the earth ran smack-dab into forces that brought our Eden to a crashing end. *The mother of all revolutions broke out against the conservative bedrock of humanity.* A perfect storm of cataclysms toppled *the rule of the best*, organically developed in our species over eons – and much of subsequent history became a prolonged *reign of terror* for our young.

Those leading the revolutionary charge were not enraged mobs of long-suffering *peasants*. They were wave after wave of enraged *parents* who answered the call of revolution by arming themselves with rods and whips to beat down their offspring on the slightest pretext. They heard not the cry for love because they were so busy heeding the rallying cry to pass on from generation to generation the revolutionary torch to gut the conservative order of humanity.

A century ago the assassination of an archduke set off a world war. Millennia ago a war of the worlds touched off endless waves of assassinations of babies as the prototype of royalty. Swept away by catastrophic events beyond their control, and starved for emotional sustenance, survivors lost track of what is owed the child as the true royal in our midst. Long before people learned to bow and scrape to cardboard kings, tricked out in gaudy trappings, they mounted a *regicidal* campaign against the real thing. Words like *perfidious* or *sacrilegious* are too weak to do justice to this turn of events. It is the radical overthrow by the most violent means of our royal birthright. In one fell swoop there occurs an earth-shattering breakdown of the reverence and homage owed the young. And parenting goes from raising children as sovereign beings in their right to dragging them kicking and screaming from the seat of love to commit crimes of *lèse-majesté* against their person – exiling them to the wild, stripping their dwelling-places of all that speaks to their worth, and targeting them with pent-up stores of murderous rage and contempt.

This is the greatest revolution of all time that nobody ever heard about! It began with an apocalyptic attack on the kingdom of earth. Marauding hosts of the Lord, over-running our terrestrial seat, left a trail of dismembered families and exploded communal support systems that children depend on to be treated in the manner they are accustomed to from the womb. The overthrow of *cosmic* order enrolls parents as firebrands to overthrow the *natural* order of humanity. The agents of this archeo-revolution are not howling mobs taking up muskets and bayonets against kings, but screaming hordes of parents who come armed with whips, switches, ferrules and canes to break the child's spirit and quash his right to be a sovereign being in his own right.

A British parliamentarian frets and fumes over the fate of tens of thousands of aristocrats in a short-lived Reign of Terror. Where is the Edmund Burke to mournfully reflect on the deaths of tens of millions of children abandoned or persecuted to death – not in a corner of Europe but the world over; not over a couple of years but over the entire course of history? To put it mildly, the impact of this revolution has yet to be fully assessed. Suffice to say, it proved more destructive than a thousand French Revolutions. By the time it raged for a few thousand years, it left an indelible blot on the face of civilization. Its leading lights didn't have to be wild-eyed zealots like Robespierre. They could be mild-mannered humanists like Montaigne, literary star of the French Renaissance. In an age where child mortality was high, there were statistical reasons for his throwaway line: "I don't count the little one."[1] But by feeding into the mindset where children do not count, it bespeaks a model of parenting that denudes children of their sacred VIP status. They may not be left to languish in isolation like Princes in the Tower, but from birth they are mummified in shrouds, put behind bars of cribs, arrested in their range of motion and emotion under the guise of keeping them out of trouble. They may not be carted in tumbrels to the jeers of the crowd, but they are pelted with smears and slurs and bombarded by the blistering criticisms of those who fancy themselves their betters. They may not be dispatched to the guillotine, but they are frog-marched to a place where they are psychologically decapitated by being cut off from the organic wisdom of their bodies and forced to retreat into their heads to survive.

With a dawning science of human relationships, you would think psychologists would inject some common sense into a culture where innocent babies fill the role that ill-fated Bourbons or Romanovs did in the mind of

---

[1] *Children Through the Ages, A History of Childhood*. Barbara Kaye Greenleaf [p. 32]

Robespierre or Lenin. Think again. Monarchy is a spent force by the 20th century, but hysterical charges of megalomania and delusions of grandeur still come thick and fast at the infant. When mental health professionals of good repute accuse a helpless little creature, totally dependent on our good will to survive, of *infantile omnipotence*, you have to wonder whose sense of reality is impaired – the baby's or the shrink's. Pediatricians have to be deeply wounded to mistake a baby's cry for love as a rallying cry to rise up and put the nascent monster in its place. Psychologists have to be severely disturbed to damn the tiny fingers that cling to them as a power grab. Rather than being revered as sovereign beings in their own right, babies continue to arouse fear and loathing as budding tyrants we have to crack down on and cut down to size. *His Majesty, the Baby* the psychoanalyst sneers, in supercilious tones.

It takes a village to raise a child, but the question is, what kind? A parochial village of blithering moral idiots – or a global village of emotionally wise sages? Rather than being celebrated by parents, teachers, magistrates, and priests for their transparency and spontaneity, their irrepressible curiosity and high spirits, the young are traditionally demonized by the equivalent of village mobs that come after them with torches and pitchforks – canes, whips, belts, and straps to violate and dominate them, break and beat them and make their life a living hell. Even after 150 years of advances in psychology, there are those who dare not utter an unkind word about oil kings or military-industrial czars with more power than god, but who are prepared to march on helpless children and rudely disabuse them of their genetic expectation to be treated as royalty.

Here is the motherlode of all double standards. We go

easy on our leaders because we are so hard on our children. We uncritically take every kind of evil dished out by our guardians because nothing our children do is judged good enough. We never get around to overthrowing the powers that be because we are busy flexing our muscles overpowering those who depend on us for everything. In the public arena we do nothing to oppose the triumph of evil, because in the private arena we do just about everything to crack down on the fruition of good in our young.

We point to people lining up overnight for the latest cyber gadget, or working themselves into a lather over the latest tidbit of celebrity gossip, or doping their kids into zombies with the collusion of court physicians, and say: *This is why we have no revolution in this country.* But the main reason is because the brunt of our revolutionary fervor is exhausted manning the barricades against the archeo-conservative order of our humanity and bringing it down any way that we can.

As humans are not cruel by nature, it is imperative to clear the record of blood-stained cruelty against the young of our species. The maltreatment that is the dominant feature of childraising for ages rests on an epic distortion of the nature of children. *Children are bashed as demonic imps of chaos, a threat to the social order, a specter of anarchy that must be vigilantly checked and suppressed, summarily punished and put in place to save the sanctity of the domestic and social order.* When a tract depicts a cabal of Old Men plotting to take over the world, it is denounced as a forgery. But the juvenile version of that myth passes unchallenged. Rooted in archaic fantasies of infantile omnipotence, in historic fears that babies are born plotting to dominate the world, it could be called ***The Protocols of the Youngers*** *of Zion*. Here is the Magna Carta of child abuse. Here is the enabling myth for soul murder and persecution in the family. It is what led Pharaoh to drown the children of the Israelites. It is what led King Herod to decree a Massacre of the Innocents. It is what continues to lead a whole school of pediatricians, psychologists and misguided parents to impute to babies "the magical quality of being cunning, power-mad schemers who must be taught discipline and self-control from the moment they are born."[1]

This is the greatest conspiracy myth of all time. As readily as we embrace conspiracy myths against children without a jot of evidence, we are loath to see all the insidious ways our guardians plot to dominate the world, even when the proof is before our eyes. We swallow fantastic theories of children as troublemakers, *enfants terrible,* while blandly dismissing as conspiracy theories all the ways in which our troubles stem from a terrible brand of leadership. *Discord is the lifeblood of all political organizations. Rather than eliminate conflict, the State must encourage and promote threats against which it can mobilize and control its own population.*[2]

A fatalistic tone of resignation runs through all our Twitter tempests in a teapot. We fail to draw the line against the relentless encroachments of those in corridors of power, even as we go overboard laying down limits for those in day care centers. We deflect blame from institutional

---

[1] "Learning From My Children" Vivian Janov. *The Feeling Child*, Arthur Janov [p. 204]

[2] Butler D. Shaffer, **Calculated Chaos, Institutional Threats to Peace and Human Survival** [p. 115]

evildoers with baseless theories of innocence,[1] while savaging innocent children with all kinds of scapegoating rituals and character assassinations. We practice political quietism with lords of finance who push us back into the Dark Ages by making it harder for each generation to do as well as the one before. We hold national conventions to ballyhoo candidates who promise a break from the past, while refusing to endorse children who represent the last best hope for the future. Our struggle with bosses who control every facet of our lives lapses into quiescence, as our battle of wills against toddlers who need to be themselves escalates into the Terrible Twos.

How can we not fight a losing battle for freedom in the real world as long as we keep tilting at imaginary ogres and phantom tyrants in our young? Those who act like resistance against institutional guardians is futile, who are convinced you cannot fight City Hall, are in the forefront of the movement to gag the still small voice of humanity, and aggressively campaign against the touchy-feely values of our mammalian heritage that are gateways to a kinder, gentler world. If instead of waging propaganda war on kids we loved them as they are, we would liberate the energy we put into suppressing our feelings and blow away the bases for political repression and economic oppression.

Society would be better off trusting its children more and its leaders less.

------

[1] Such as "coincidence theory" or "somnambulist theory" or "idiosyncrasy theory" or "incompetence theory" or "aberration theory" and the like. *Land of Idols, Political Mythology in America*, Michael Parenti [pp. 160-164)

26. WHEN MANKIND
BECOMES A FAILED STATE

When man stole fire from heaven, it wan't to ward off wild animals or roast meat. It was to ward off the devouring wrath of gods by making burnt offerings of children. To offend divine or human parents wasn't *like* the end of the world – it *was* the end of the world. To avert this dreadful possibility the most precious thing in the universe is thrown away – the newborn baby.

It still makes headlines. "Newborn Baby Found in Garbage Can."[1] The Probable Cause statement filed by police stated the birth mother was afraid to tell her parents about her pregnancy and discarded the baby in hopes it would die and solve her problems. What's with children and their preternatural fear of upsetting parents? Our ancestors lived in dread of the judgment of gods, and damn, if infanticide didn't figure in their solution too! If they had TV cameras in those days, we could see footage of mothers dropping their babies in burning pits in the Valley of Hinnom, a site outside Jerusalem, in hopes of staying on the good side of their Father in heaven. From *Hinnom* or *Gehinnom* comes *Gehenna*, the biblical word for hell. In medieval times it became a place for burning rubbish where the flame never went out. Can we spell it any clearer? *Children = garbage*. Trashing babies gives a new lease to the theory and practice of hell. While no one knew when God would return to judge man, generations of rambunctious children were warned on a daily basis to wait for their fathers to come home and give them hell. And who

---

[1] Marcos Ortiz. ABC Channel 4 Utah. Good4Utah.com

can blame these battered victims for growing up to get a belated revenge by damning others to hell.

Hell as a human garbage pit of eternal fire is a smoldering relic of Doomsday. A scientific age would have chalked up Doomsday to a cosmic accident. A prescientific age took it for a fiery judgment passed on a rotten race. Not Doomsday itself but the religious spin on it doomed mankind to be judged in perpetuity. For man to be branded as evil and punished by the rod of *divine* wrath set the precedent for trashing the innocent children of men as evil and thrashing them with the rod of *parental* wrath. One has only to gaze up at the painted ceiling of the Sistine Chapel to see how the oar of Charon, upraised against sinners in the Last Judgment, could easily have doubled as a weapon to beat back generations of young into a hell of torment and suffering.

Today a mother who stuffs her baby down a trash chute, or dumps it in a toilet, is a scandalous object of horror, pity and blame. In days of old, such an unnatural mother performed a public service. She averted an ominous threat to the social order posed by the newborn baby. At the same time, her unspeakable act served as a rite of passage for incubating the movers and shakers of tribal culture, the leading lights of civilization. "Child abandonment is a universal motif in myths. The culture hero is often a child who is thought to have been killed, but is then somehow miraculously saved. The child then often fulfills a predestined fate. Examples of this are Moses, Buddha, Jesus and Oedipus."[1] Come again? Granted that the catastrophes that

---

[1] *Kalevala Mythology, Revised Edition* [p.213] Juha Y. Pentikainen and Ritva Maarit Poom

laid waste to earth led to treating our young as waste matter. But how does it midwife the birth of a pantheon of culture heroes for children to go from getting the royal treatment to being given the bum's rush and treated like crap? Why should a child god be an abandoned foundling?[1] The tip-off is a little thing that could be christened the *paradise paradox*. The blissful home is literally a walled enclosure known as paradise. In one sense, it's the height of odium to be *eliminated* from it – to be forbidden to darken its threshold (*limen*) again. In another sense, it's the path of apotheosis, for one cannot be driven out from this walled enclosure without transcending the surfaces and sides that normally circumscribe one's existence and taking on the boundless aspect of a god.

Cosmic order is the capstone of existential security. Suns rise and set. Stars glide in their orbits. Earth abides under our feet. "Now imagine that natural phenomena should suddenly begin to contravene the laws of physics – that is, that we should lose confidence in science...the faith by which modern European man lives."[2] We face a crazy world that flies in the face of rational expectations. The French Revolution had its Reign of Terror. *The apocalyptic revolution our ancestors faced had its reign of existential terror.* "The house in which we had installed ourselves would fall in. *In material matters we would no longer know what we could **trust***; and the human being would again be cursed by that terrible blight which for thousands of years

---

[1] ***Essays on a Science of Mythology***, C.G.Jung and C. Kerényi [p. 27]

[2] Jose Ortega Y Gasset, ***Man And Crisis*** [p. 37]

caught him by the throat and held him prisoner – that cosmic horror, that panic terror, the fear of **Pan**."[1]

The toppling of our terrestrial house leads to the collapse of our domestic home. The effect of mother earth and father heaven being no longer there ripples out until mother and father are longer there. As our maternal universe collapses like the material one, in *maternal* matters we no longer know whom we can trust – unless it be our self! Here is the origin of the foundling as child god. With no one to rely on save ourselves, self-reliance becomes the watchword of existence. And we are called on to pull off the signature feat of culture heroism – miraculously lift ourselves out of the swamps of nothingness by our hair to become the ground of our own being.

The desolate worldview that molds the royal pretensions of the culture hero thus goes flat against the norms of humanity. When a birthing mother squats, the baby pushed out of the birth canal has an innate expectation of someone waiting to catch him. It is not in our genetic programming to be pulled down by gravity and allowed to crash. Someone to hold us, cradle us, support us, is key to our development. Without that experience, trust is impaired and we are at risk of joining the ranks of the "bondless men, women and children (who) constitute one of the largest aberrant populations in the world today."[2] A belated recognition of this risk gives rise to the Trust Fall game, where adults seek to recapture the primal trust lost in childhood. A volunteer falls backwards into a group,

---

[1] Ibid.

[2] **High Risk, Children Without A Conscience**, Dr. Ken Magid, Carole A. McKelvey [p. 57]

trusting to be caught as he falls. The point is we need to know we are not alone. *Others will catch and protect me from hitting the ground and hurting myself.*

Trust-building is an exercise in cultivating dignity and worth. A human being is like a priceless vase. How many of us can stand by and watch a porcelain vase totter and topple? Are we not impelled to rush forward to catch it? The greater the vase's worth, the more desperately we lunge to save it. But if the vase is a cheap copy, we can let gravity take its course. It is no great loss if it falls and smashes to bits. We sweep up the fragments and toss them in the trash.

The implications are far-reaching. With no one to catch and hold me, my trust in others is shattered. My self-worth and dignity take a devastating hit. I am not just liable to crash and smash a bone. I am liable to fall and break my heart. If the lesson is driven home on an existential level, my being is trashed and I'm haunted by my nothingness. "Imagine the ultimate terror of not being. There are people who do not have a sense of their own existence, there is a deadness within."[1] This inner deadness turns me loose like a vengeful ghost, bent on haunting the world from a place where I am no longer dragged down by the reality of my flesh and blood.

The beauty of being enfolded in a state of closeness is the closure it provides. *So long as I'm here, this is where you belong.* When the closure of that embrace is missing, one suffers a boundless crisis of confidence and there is nothing to hold one back from going to the ends of the earth in a frantic quest for territories and markets to scoop up into one's all-encompassing arms. Denied loving arms to

---

[1] Ibid. [p. 64]

respond to their needs, babies might just get mad enough to take up military arms to make the world respond to their whims. Bereft of their birthright of homage and adoration from the tips of fingers, babies just might be tempted to wrest it at the point of a gun.

When the world fails to respond to my needs, I survive by becoming responsible for the world. With no one to rely upon outside of myself, I become like the frontispiece of Hobbes's **Leviathan**, with a world of little people incorporated into myself. With no one to catch me, my trust in the world is shattered as I fall – *and when I fall, it all falls on me! The weight of responsibility for the world falls on my shoulders.* This is the origin of the *imperial ego* enshrined in the institution of kingship and autocracy. Abandoned as a child, I defend against my fear of being swallowed up by a pit of *nihility* through a mandate for *omneity*. Psychohistorians point out the "eerie balance between… suicidal Nothingness and dictatorial Allness."[1] As a lack of love turns the heart to stone, the link between Nothingness and Allness is inscribed in hearts of stone.

Teen gangs vandalize entire neighborhoods to proclaim to the world, *I am here.*[2] Orphan twins, abandoned to be raised by wolves, cope with the terror of nothingness by becoming the founding fathers of one of history's greatest empires. We glibly mouth the phrase *Aut Caesar aut nihil* without psyching out the link between Caesar and nihility.

---

[1] Erik H. Erikson, **Young Man Luther, A Study in Psychoanalysis and History** [pp. 107-108]

[2] "By far the most terrible thing I learned," she says, "is that assaulting, harming or killing others is a way to say, 'I am here.' Nelsa Curbelo, an.activist who turned a gang-infested city in Ecuador into *Barrio la Paz* ("Peace Town")

The sheer terror of being *nothing* impels us to be *all. Love is everything to mankind. Without it there is only the hunger for everything.* And what we have then is a fanatical quest to fulfill, not the needs of mankind, but the all-consuming ravenings of a fantastical monstrosity that could be mockingly called *pankind.*

27. THE TRASHCAN BABY AS
THE FATHER OF EXISTENTIALISM

What comes to mind when we think of 'conservatism'? A traditional system of rule? A time-honored if not hidebound worldview? A beacon of order, a bastion of stability? It never crosses our mind that it is a product of the most sweeping revolution of all – the ur-revolution that destabilized the planet and wrenched us loose from our moorings in the human continuum.

Conservatism as we know it is the motherlode of organized conflict, chaos, and crime, because its vaunted elements – pseudo-stability, pseudo-security, pseudo-law, pseudo order – derive from a universal chaos, a catastrophic breakdown of cosmic law and order. Its characteristic features – kingship, autocracy, imperialism – all come out of a disastrous shattering of trust, a collapse of the trusted supports of our existence. Conservatism has blighted man's hopes for peace and security for thousands of years. It has caught man by the throat and held him hostage to its exorbitant demands. Its guiding principle is that nothing less than All can satisfy man and save him from a cosmic terror, a horror of Nothingness. And so it is gripped by a rage to push the bounds of acquisition, to stretch the boundaries of conquest, to indulge an all-consuming hunger, a panicked drive to approximate as nearly as possible that *Allness* that is the literal meaning of *Pan*.

The philosophy that best captures the spirit of conservatism from the dawn of history is that most modern of philosophies – *existentialism*. As a philosophy, existentialism has the symptoms of Trashcan Baby written all over it. *Abandonment. Anguish. Dread. Despair.* In this desolate state, wrenched from the maternal universe, one is stabbed to the quick by the specter of one's own nothingness. The contours of one's nature disintegrate. The ground of one's being gives away. And it is in this solitary state, this dreadful abyss of isolation, where there is no one to depend on outside oneself, that one has a complete and profound responsibility over the world.

It is not a philosophy for most of us, thank goodness. It is the working philosophy of a fringe bent on world domination. What kind of people presume to lord it over their kind, cut off from the rules of humanity, the exchange of fellow-feeling, and the give and take of community? People who view life through the bleak, scary lens of the Trashcan Baby. It is not a world committed to making our stay on earth a long happy one, but one bearing the apocalyptic imprint of infanticide – "a blackness which no stars redeem, a wall of terror in a night of cold."[1] Being unceremoniously dumped at birth makes it official. No refuge is to be found in another's heart. No one takes pity on or pleasure in our existence. **The world is an inhospitable place out to get us.**

As the other withdraws on all sides, what is the self supposed to do with the vacated space around it? You think nature abhors a vacuum? Check out the ego in an emotional vacuum. *It expands like crazy to take up all the space around it.* Bereft of the other's defining presence, it pushes back its boundaries as if others don't exist. Instead of healthy growth, the self undergoes *hypertrophy* or abnormal growth into a giant or god. Instead of becoming magnanimous, it is overcome by *megalomania*. And instead of fine-tuning the system of checks and balances between the

---

[1] Robert Bridges (1844-1930) from the sonnet "Melancholia"

self and other to make life grand for the both of them, the self succumbs to delusions of grandeur and aggrandizes its claims to the point of dwarfing everybody else's!

The Trashcan Baby is father to a class of egos that are so boundless they crowd everyone else out of the picture and wipe them off the map. Existentialism opens the ground of human nature under one's feet. "I cannot base my confidence upon human goodness or upon man's interest in the good of society, seeing that man is free and there is no human nature which I can take as foundational."[1] Denatured freedom is the jumping-off point for existentialism – a springboard for transcending the norms of humanity and achieving that *superhero* status or godlike *supremacy* where there are no others beside oneself, and one is accountable to no other but oneself.

The anguish and dread of the existentialist baby comes from a shattering of trust. Most of us who live normal lives are strangers to that stark horror. Somewhere in our lives we find a thread of reassurance: "If you're lost you can look and you will find me/ Time after time/. *If you fall I will catch you, I will be waiting/ Time after time.*"[2] For some of us, this is just an empty promise that turns into an act of treachery, a savage betrayal, by becoming the defining story of our lives. The picture is that of a father who holds out his arms to catch the toddler – and when the toddler obligingly takes him up on his invitation, lets him crash to

---

[1] Jean-Paul Sartre "Existentialism is a Humanism" *Existentialism from Dostoevsky to Sartre, ed. Walter Kaufmann* [p. 299] Meridian Books

[2] Cyndi Lauper, "Time After Time" (1983)

floor. The father then takes a sadistic glee in driving home the moral: *Never trust anyone completely, not even me.*[1]

Imagine a childhood harrowed by such betrayals...a father who is not a protector but a con artist, bragging of *cheating* his children every chance he gets to make them '*sharp*'. Would it surprise you to know that such a child grows up with a disastrous mistrust of his fellow man – and only feels safe vacuuming up all the resources, all the money, all the power he can to become an oil magnate, a titan of industry, a banking colossus, widely hailed as the world's richest man, whose name, John D. Rockefeller, is synonymous with the obscene wealth of a modern Croesus?

The trust-shattering anguish, despair and abandonment of the existentialist plight "puts every man in possession of himself as he is, and places the entire responsibility for his existence upon this own shoulders."[2] Doesn't this give rise to the mantrum of self-reliance that is the keystone of a 'conservative' philosophy? By pulling myself up by my bootstraps from the nothingness of my being, I become the poster boy for the bootstrapping hero who plays a starring role in the mythology of 'conservatism', primarily because it justifies the destitution of the have-nots as being good for the soul. If there is no one to catch *me* when I was growing up, why should I care if there is no one to catch *you*! Doesn't the cult of self-reliance rationalize the cruel, callous policies that shred the *safety net* that people depend on to catch them in a dog-eat-dog society, where

---

[1] *John D. Rockefeller, Anointed With Oil*, Grant Segall [p. 15]. The quote in the sentence that follows [p. 15-16]

[2] Sartre, "Existentialism is a Humanism" [p. 291] The reference to fashioning man in the next paragraph [p. 292]

THE REPARENTING REVOLUTION

they are at risk of falling below the poverty line, slipping into ill-health or homelessness, or even sinking into the grave because they are denied rightful access to the necessities of life?

If existence precedes essence, and if I am nothing to begin with, doesn't that feed into a glorification of the man who *starts out with nothing* and by dint of hard work and initiative builds himself into a *self-made man*? "I am thus responsible for myself and for all men, and I am creating a certain image of man as I would have him to be. In fashioning myself I fashion man."

I fashion man? Now that should make your skin crawl! For it is the goal of a eugenics philosophy that treats human nature as a fiction, a lump of inchoate matter which one is free to sculpt into the superman of one's dreams.[1] Which is why Rockefeller, whose 'conservative' platform made him the apotheosis of the 'self-made' man, was drawn to fund a eugenics movement that inspired the master race theories of Adolf Hitler and his ambitious project to make man anew.

---

[1] Hannah Arendt, **Totalitarianism** [p. 412]

28. TRADITIONAL CONSERVATISM – THE IDEOLOGY OF THE TRASHCAN BABY

Words like *discipline, boundaries, limits* loom large in a 'conservative' vocabulary. But it is mostly for others to practice, not for those who preach them as articles of faith. *Power is the rule of hypocrisy. It is the freedom to do what I want, while others are compelled to do as I say.* It is what makes 'conservatism' so fake.

Power corrupts through a rupture of trust. With the shattering realization the other is not there, there are no meaningful boundaries to demarcate where the self ends and other begins. Existentialism construes this lack of boundaries as *God is dead, everything is permitted.* What God is believed to be for man, love is to a child – a humanizing presence, a bonding mechanism that socializes us in the art of co-living. When the light of love goes out, our eyes are too benighted to appreciate the contours of our being. Feelings numb out, consciousness dims, integrity becomes a blur. When *love* is dead, everything is permissible.

The reality of being unloved is we can die. The morality of being unloved is we can no longer live and let live. We cope by taking matters into our hands and looking out for number one. We become monopolists when the world cannot be trusted to meet our needs. We corner the market when others cannot be counted on to share. Carrots and sticks are necessary when the human connection withers. Carrots buy obedience and sticks enforce it. Militarism becomes a tool of coercion and money goes from being a medium of exchange to a means of enslavement.

Out of the failure of trust comes *paranoia.* And out of paranoia comes *megalomania.* The obsessive watchword of paranoia is *the world is out to get me.* The motive force for megalomania is the compulsive reaction to reach out and *get as much of the world as I possibly can.*

Parents who insist on free rein to exercise authority turn out the most rigid and repressed children. The Convention on the Rights of the Child is the Magna Carta of our young, spelling out their political, economic, social, and cultural rights. The U.S. is the only industrialized country to refuse to ratify this U.N. treaty because 'conservatives' in Congress oppose it. *They fear it places undue limits on the authority of parents to do as they like with their children!* You get the big picture? Disciplinarians are overly strict at setting boundaries for children because so-called conservatives are much too lax at setting boundaries for parents! If to be spoiled is to spurn limits and boundaries, who are the spoiled brats here? Children who expect to be safeguarded from parental abuses – or parents who chafe at rules and regulations because how they run their households is their business. Now if that doesn't just beat all as a *free market version of childraising!*

Along the same lines, when the existentialist crows, *Man is free, man is freedom,*[1] it is not necessarily something to rejoice about. It is not the freedom *of* humanity to be what it is naturally meant to be – it is freedom *from* the essential qualities that define human nature, like kindness, sensitivity, vulnerability, empathy. It is not the freedom of man to be himself – it is the freedom of man to wrench himself from the limitations of his psychobiology and act as if he were *god!*

This pseudo-freedom is the bane of man from the dawn

---

[1] Jean-Paul Sartre. ***Existentialism from Dostoevsky to Sartre***, ed. Walter Kaufmann [p. 295]

of religion! Man cannot aspire to a more than human or *superhuman* freedom without damning his fellows to a less than human or *subhuman* servitude. By denying the centrality of human nature and knocking it out of the equation, existentialism gives credence to a world made up of gods and beasts, supermen and submen. Sartre extends the line of continuity from Nietzsche to Hitler to its logical conclusion. His philosophy condemns man to be free, but free to do what? To *impose a hubristic rule that runs the gamut from theocracy to totalitarianism*! First it gave rise to royal pretenders known as priests and kings, who made life wretched for people by lording it over them as gods. When they go out of style, it gives rise to a class of economic royalists known as bankers and industrialists, who make life no less wretched for people by reducing them to wage slavery and debt servitude.

Can a philosophy forged in a cauldron of anguish and despair treat the pursuit of happiness as a sacred right? The power to expand is normally limited by how it impinges on, impairs, or diminishes the existence of others. The agonizing collapse of trust from a calamitous absence of others hurls one into an abyss of self-reliance that precludes the need to consider others. It is the ideal crucible for the apotheosis of the *imperial ego*. For as long as the drive to amass wealth or power is unimpeded, there are no limits on how big or absolute one becomes. Neighborhoods and communities no longer mean anything. The environment no longer matters. The integral needs of human nature are of no more account than the ecological demands of terrestrial nature.

All the lip service existentialism pays to freedom cannot hide its ulterior agenda of totalitarian control. A culture hero, ripped from a web of social relations, has the arrogance to impose a whole set of factitious, if not fact-challenged rules on society. Out of the anguish of abandonment the existentialist has the audacity to become the type of *legislator deciding for the whole of mankind.*[1] Where does he get off as lawgiver or policy-maker for us all? The answer stares us in the face. Unable to trust man to respond instinctively to man, the existentialist becomes *responsible for all men.* And what better way to put his money where his mouth is than with a system of untrammeled capital accumulation that makes *his* the go-to philosophy of a capitalist elite that reduces democracy to a running joke by turning its public servants into a retinue of private ones.

The place to see existentialism in action is the free market. The free marketeer is an existentialist who wages a vendetta against rules and regulations designed to curb his excesses, thus usurping the wannabe-god-given right "to act in such a manner that humanity **regulates** itself by what I do."[2] It takes a world-class narcissistic personality to shake off all regulations in the same breath that it decrees that everyone else is bound to be regulated by what it does! Hailed as an anthem to freedom after the nightmare of fascism, existentialism is every dictator's wet dream.

The double standard is the double strand in the DNA of existentialism. If there are no essential rules to define my existential freedom as a man, I am free to run amok and play god. What good can come of this double standard for

---

[1] Jean-Paul Sartre, **Existentialism From Dostoevsky to Sartre**, ed. Walter Kaufmann [p. 292]

[2] Jean-Paul Sartre. **Ibid.** [p. 293]

humanity at large? People are dragged off to prison so that prices can be free.[1] Lives are reduced to debt servitude so that markets can be free.

Behold, the hubris of *conservatism* on display in all its hypocritical glory! Let the culture hero as lawgiver or legislator write the laws of the tribe. Money masters go one better. *Give me power to control a nation's money and I care not who writes its laws.* It's no great loss to abdicate the divine right of kings when you arrogate the right as Chairman of the Federal Reserve to compel the world to fix its eyes on what you say and do and regulate its behavior accordingly![2]

It sounds grand when existentialism proclaims, "In choosing for himself, man chooses for all men." Never mind how toxic it is to *society's* health when the lords of finance pay off Congress to outlaw all kinds of healing modalities – including those dating back millennia – in favor of an allopathic model of medicine that serves as a procrustean bed, or rather, *sickbed* for man. Or how toxic it is to our *planet's* health when the gods of oil take a wrecking ball to alternative forms of transportation to fossilize an engine fueled by their poison of choice. Existentialism is a fitting philosophy for a civilization raised in the shadow of a monotheistic god beside whom no others exist. Once this jealous god gains a new lease on life as the Almighty Dollar, there are no lengths to which his standard bearer is not ready to go to enforce obedience to the latest commandment handed down, if

---

[1] Eduardo Galeano, 1990. **The Shock Doctrine**, Naomi Klein [p.144]

[2] Jean-Paul Sartre. **Existentialism from Dostoevsky to Sartre** [p. 293] The quote in the next line [p. 291]

not from the mountain top then from the peak of the money pyramid: "Competition is sin."

Some say the Man behind the Curtain is an Old Man – others, a cabal of world-dominating Elders. No one tells us it is a broken child – a Trashcan Baby birthed in a shattering betrayal of trust. A Trust Fall would reverse the damaging effects of betrayal in the presence of those who have our backs - but it reckons without the *Betrayal Fall*. This is where a child is betrayed here and now to spare him future betrayals. Beware of parents who, under a pretext of preparing a baby for a dog-eat-dog world, invite attack dogs into the home to maul it. You cannot protect a child from those who may *potentially* abuse his trust by becoming one of those who *actually* do.

That has never stopped our guardians from making a great show of protecting society from predators by becoming the *predatory class par excellence*. And it does not stop the Betrayal Fall invoked by *John D's* father from being invoked yet again by the father of a boy named *Roger E*. "The cruelest lesson Roger would speak of occurred in the bedroom Roger shared with his brothers. Roger was standing on the top bunk. His father opened his arms wide and smiled. 'Jump Roger, *jump*," he told him. Roger leapt off the bed into the air towards his arms. Robert took a step back. His son fell flat onto the floor. As he looked up, Robert leaned down and picked him up. 'Don't ever trust anybody,' he said."[1]

When firemen urge a trapped child to jump from a burning building, and then unceremoniously yank his life net away, he takes flight from the world in a mist of blood

---

[1] Gabriel Sherman, ***The Loudest Voice in the Room*** [p. 7]

splatter. When parents do something similar, the child copes with his shattered trust in reality by taking flight into a world of make-believe and wishful thinking. He may grow up to become the founder, not of a political empire like Romulus, but of an empire of delusions, fantasies and lies like Roger E – Roger Eugene Ailes, that is, chairman and CEO of the universe of alternative facts known as Fox News.

A schizoid break from humanity often goes hand in hand with a schizoid break from reality. Which of course begs the question. If the collapse of the world gives rise to the imperial ego, how does it make sense for the imperial ego to be in charge of world order? The desolating withdrawal of human touch, the devastating eclipse of human sensibilities, the disastrous break from a consensual reality, gives rise to a *hallucinatory bubble*. And a good thing too, for the *imperial ego* only survives inside an *hallucinatory bubble*! The march from divine emperors to imperialist lords of finance requires not a progressive grounding in reality but a "breaking new metaphysical ground in the mechanics of untruth."[1]

From demigods as foundlings to lords of finance who owe their overreaching delusions of grandeur to a calamitous failure in bonding, a disastrous breakdown of belonging, *conservatism* is the ideology of the Trashcan Baby. *Talk of one man's trash being another's treasure!* Those who trash our dreams of being one big happy family are nothing if not consistent over the centuries. They have always embodied the megalomaniac isolation, the paranoid collapse

---

[1] Matt Taibbi, "The End of Facts", **Rolling Stone** February 23-March 9, 2017 [p. 28]

of trust that is the curse of the baby that comes as close as possible to being contemptuously left out to be picked up like a piece of trash. And when all is said and done, what is recorded history if not the narrative of a civilization that treasures the Trashcan Baby as God's gift to humankind.

IV. PSYCHOLOGY OF DEEP SPACE – THE UNDERBELLY OF PROGRESS

"The best leader isn't the biggest or the strongest. The best leader is the one who cares the most."

— ***Barnyard: The Original Party Animal*** (2006)

29. MAKING EACH OTHER HAPPY IS WHAT IT MEANS TO BE HUMAN

The motto of progress is the best is yet to come. Technicians work to invent a better cell phone. Geneticists work to produce a better strain of rice. Activists organize to create a better world. According to this school of thought, the lucky ones are the Johnny-come-latelies who show up at the *futuristic* end of the timeline. The best of progress is reserved for those not yet conceived or born. They are the ones to reap the greatest fruits of man's unflagging ingenuity and industry, benefiting from high-tech breakthroughs we can only dream of. They are the ones for whom the gentler world we are in the process of creating will come to fruition. The rest of us have to make do with a *work in progress*.

How unfair is that? It's like saying only those born *after* the advent of the messiah are saved. Is science the new messiah that bifurcates history into a before and after? Are those living in a prescientific age doomed to languish in a world where colors are less vivid, sounds are less vibrant, touch is less thrilling, taste buds are less sensitive? Are the receptors for pain and pleasure not as acute in those who lack the scientific smarts? Is birth less of a joyous event, or death a sad one, in BC times or AD ones?

And what about the capacity for enjoyment. Is it less developed in those who live life at a leisurely pace rather than a fast-paced one? Just because people lived before the internal combustion engine, have they missed the boat – or the bus, the airplane, or automobile? Are the great intellectual breakthroughs and technological wonders galore reserved only for those who came after the Renaissance or Industrial

Revolution? The later we are born, the more blessed we are with the latest and greatest! Is that how it goes?

There are other flaws in this argument. What about the disastrous price we force our planet to pay by breeding this or that invention beyond its natural limits? Look at the problems caused by a population explosion of internal combustion engines – the congested highways of our transportation system and the congested airways of our respiratory system. In a money-mad zeal to bring good things to light, are we not dimming the prospects of life on this planet? In a power-mad zeal to dominate the world and monopolize its resources, are we not at risk of leaving our children with a world that is frighteningly worse off than ours, where breathable air and drinkable water, let alone a life of ease and plenty, are in short supply or not even to be had by future generations? Even the invention of the light bulb is a mixed blessing when we labor under an economic system which is not above darkening the sun and the moon in order to boost sales of artificial lighting.

Much of progress makes man's future less, not more secure. Could it be the fate of our species hangs in doubt because we ignore a seminal reality? *The world of our dreams is not meant to be in our future. It was always meant to be in our past – or at least, our present.* So much so that if our potential for bliss as a species is not brought to life in our past and present, it will never be in our future. That better world we knock ourselves out creating will never come to pass as long as the truth eludes us. *The best of all possible worlds is something that each and every one of us should have here and now!*

We keep thinking it will get better by and by. What if we have it wrong? What if the perfected state we busily aspire to is destined to come not at the latest but the earliest time? And what if it requires no special ingenuity or hard work to invent or create, because all its many-splendid possibilities are handed to us on a silver platter? When it comes to material stages of development, man goes from rubbing flints to harnessing steam to pressing a remote button for a gas-powered fireplace. *What if man were gifted with a marvelous technology to create paradise on earth from day one? What if we are possessed of the means to create a life of fantastic ease and joy for each other, if not from time immemorial then at least from the beginning of our evolution as a species on earth?*

Take it a step further. What if the talent for creating a better life is not confined to a few individual geniuses like James Watt or Thomas Edison. What if, from the moment we became homo sapiens, each and every one of us is gifted with a miraculous capacity to create bliss for members of our kind. *As a matter of fact, this capacity to make each other happy is primarily what it means to be human!* Wow! No need to jump through the flaming hoops of the ages. We can create the best of all possible worlds for every generation of babies just by passing their heads through the ring of fire! Just by harnessing our emotional intelligence we can light up our lives, draw closer together, make our lives easier, and send each other to the moon. *We don't have to build interstellar rockets to reach heaven. We can just reach out and touch our beloved's face.* Imagine the pinnacle of progress where humans have everything they need to be secure, contented, joyous and at ease. That is the ideal world it has always been in our power to create! For what defines us as a species is not our brain or our ability to walk on two legs or invent the wheel. *It is our unique capacity to complete each other and make each other happy.*

30. SPACE CENTER FOR ACHIEVING HEAVEN ON EARTH

Time was when two bicycle mechanics rigged a contraption to skim the earth's surface for all of 12 seconds. Decades later visitors to the Kennedy Space Center have their teeth rattled by a Saturn V rocket lifting off into space. Impressive? Yes. A lot of sound and fury for nothing? That too. Strapping astronauts in a tiny capsule and harnessing thousands of tons of pressure to catapult them …where is it to again? The abode of the gods? A reunion of loved ones? A deathless paradise of beauty and bounty? No, no. Nothing like that. Just a stark inhospitable vacuum on the fringes of space. Or perhaps – if you were one of the sacred twelve who supposedly touched down there – a cratered expanse of a moon, incapable of supporting life even in its most primitive form.

Doesn't Mother Nature have us beat in that department? Didn't she engineer a capsule of her own – marvelously cushioned and equipped with an incredibly complex life support system? A capsule that reprises a journey through eons of evolutionary time and space in less time than it takes for the earth to orbit the sun. A capsule designed to bring us to the most beautiful planet in the solar system. There we are, weightlessly floating, all curled up and bound to the mother ship with an umbilical cord. When we arrive at our destination, it is not a cold hostile vacuum misnamed heaven. It is heaven for *real* – an intimate abode of bliss that flows with milk and honey – a place of beautiful communion and beatific completion where we come to rest in tranquil joy and content.

The space capsule in which our young make a nine-month-long journey to heaven is known as the womb.

When they disembark, they are greeted with much fanfare and made to feel welcome. They are immediately enfolded in blissful communion with the Mother of their being. They are reverently held up high in the euphoric presence of the Father of their being. Their innocence is a source of awed delight, their beauty an occasion of joy and admiration. They are invited to feast on the milk and honey of paradise.

The Space Centers named after dead presidents in Houston and near Orlando lift us into a primal void that is preternaturally inhospitable to life. What are some of the words that apply to this "heaven"? *Cold. Empty. Dark. Remote. Hostile.* These are the same words that apply to a personality type that causes all the havoc and heartache on earth. These are the same adjectives used to describe a set of behavioral traits that include a ruthless contempt for others, delusions of grandeur, intermittent explosive disorder and a sickening capacity for violence and destruction. How bizarre is it that we should know as *heaven* an expanse of the universe whose physical qualities, when personified in a human being, are a source of abiding torment and despair worthy of *hell*!

There is only one Space Center that offers the kind of heaven we are programmed to enjoy from birth. One Space Center that embodies qualities like *warmth and fullness, closeness* and *support* that are genuinely heavenly, in that they are at the opposite remove from a physical experience of the heavens. And that is the Space Center of the human body. Millions of years in the making, it is exquisitely engineered to satisfy our biological hunger for attachment, trust, comfort, security. To be taken to the bosom of this heaven sets the bar for our well-being. It is the great homecoming of the species. Our senses are keenly oriented to this Space Center from birth. Its natural rhythms set the organic tempo of our existence. Its tactile bonds lay down a lifetime foundation of sympathy and trust. Its relaxed aura of attention and acceptance, approval and affection fine-tunes the psychobiology of joy. Its emotional intelligence synthesizes the biochemistry of peace.

How can any celestial mothership measure up to the life-giving, nurturing, protective energy field that radiates from the human body – much less the incredibly sophisticated vehicle built into the female body for transporting our young to heaven? Be it *inside* the body or *out*, cocooned in the womb, or enfolded in the bosom of communal bliss, it is right there at the beginning of history – right there at the beginning of human life. Even if we could, we couldn't ask for anything more. We have it made from the get-go!

Today, there are companies that offer the novelty of a trip into space – a space adventure on which multimillionaires get to eyeball our planet from 100 kilometers out in space. Well, what if there were a company that offers the experience – not of how beautiful our planet earth looks from space – *but of how beautiful life on earth can be right here on the ground*? What if there were a company that offered to take us to a paradise of communal rapport and indescribable joy – a place of peak joy and bliss, not to mention deep understanding, appreciation, inner contentment. What if it had the technology to offer that kind of experience? Everyone would be banging down the doors. If the going rate for space tourism is any indication, one suspects the price would be stratospheric.

As it turns out, this futuristic company has been around for thousands of years. We don't have to ante up millions for one of its tickets to paradise. A complimentary ticket is held for us at will-call. All we have to do to get our hot little hands on it is shove our way through the birth canal. For it is the *company* of our mother and our father.

31. HEAVEN AS THE PARTICIPATION TROPHY OF THE AGES

Scrooges of 'conservatism' grouse at the baby boomer practice of awarding participation trophies. Handing out trophies to everyone, regardless of achievement, promotes the idea that *everyone's a winner* (gasp!) Clearly someone never got the memo. The genuine conservative order of our being – our beauty and innocence, our capacity for feeling and fellow feeling – flowers under a shower of unconditional love. Love is the greatest participation trophy of all time. Babies bring no effort or skills to stake out the milk and honey that makes them the happiest creatures on earth. To run a tab on what they imbibe from the fountain does a number on the conservatism at the heart of their being. The candle of life cannot be expected to shine unless it feeds on the oxygen of love. To flip that on its head, by claiming the oxygen of love is some sort of a prize that only deserves to be awarded to those who shine, is to go against science and common sense.

Children are the first to grasp the wisdom of unconditional love and the last to quibble with it. But not when they are outfitted with lenses that distort their vision and leave them myopic. Such is the case of the 12-year old whose views are featured in a NY Times op-ed.[1] Someone ought to sit this poor girl down and give her a good talking-to. *Yes, Betty, it's in our DNA to expect awards for just showing up. What's wrong with that? Isn't that what you expected when you showed up in a maternity ward? What did you do*

---

[1] Betty Berdan, "Participation Trophies Send a Dangerous Message" **New York Times,** Oct. 6, 2016

to deserve being bundled up and whisked off to be raised in the affluence of a middle-class home? What did you do to merit a nutritious diet? The best pediatric care? Or to be sent off to the best schools? What games did you excel at to be rewarded with braces, ballet lessons and birthday parties? What championship prize money did you win to cover the $233,610 it costs to raise a child?[1] Was anyone even keeping score when the horn of plenty fell in your lap because you happened to show up through a dilated cervix?

Scrooges have reason to be troubled by the awkward implications of participation trophies. If awards are going to be handed out like candy, how come a girl in the Congo is scrounging for discarded KitKat wrappers to lick and a girl in Connecticut has a bag bulging with Halloween treats? If we didn't know better, we might suspect Betty Berdan to be interning for the likes of policy-maker George Kennan and writing a junior high school version of his infamous Memo PPS23. You know, the one where he observes that Americans are only 6% of the world population, and yet enjoy over 50% of its resources. When only a select minority can belong, we can't let sentimentalism and idealism cloud our judgment. We can't be enlarging the circle to invite the rest of the world to partake of the good life. Our elitist position depends on circling the wagons and drawing the winner's circle tight.

O power – hypocrisy is thy middle name! Funny how *faux* conservatives never grumble about all the mediocre people who get to be Presidents and CEOs because their rich, powerful daddies pulled the right strings! When was

---

[1] "It costs $233,610 to raise a child", Kathryn Vasel CNN Money January 9, 2017

the last time you heard a faux conservative griping about the dangerous message sent by the Walton family? What did they do to deserve being the richest family on earth? Is all their money less valuable because it piled up through an accident of birth? Is their perennial place atop the Forbes list of the world's richest people less meaningful because they did diddly-squat to achieve it?

Only pettifoggers begrudge kids a medal for teaming up to kick a ball around. If they lifted up their eyes they would see an entire firmament of stars and planets, choreographed to dance in providential harmony above our heads through no doing of our own. And what about a religion that coaches human beings to forego food and female companionship, to flog themselves and endure herculean austerities, to be worthy of a make-believe place in a celestial podium of the elect when they die? Didn't anyone tell them the earth already occupies the winner's circle in the heavens – perfectly pirouetting in place to enjoy sun and rain, and flourish with everything necessary for an abundant life? Did any athlete of God ever have to lift a finger to shoot the earth into the Goldilocks hoop?

When it comes to heaven, we can get the facts prosaically from a cover feature in a popular news magazine: "The urge for heaven is universal; we need it the way we need love."[1] Or we can trust a Romantic poet like Lord Byron to cut through the metaphysical gobbledygook and get to the heart of the matter: "Love is heaven, and heaven is love."

The beauty of it is the heaven-as-love paradigm is our birthright. It is owed us just for being born. Love has no conditions to meet because love is the precondition of life.

---

[1] *Newsweek*, August 12, 2002, "Visions of Heaven", Lisa Miller [p. 47]

So many of us turn life into an extended science project where we keep having to *prove* ourselves to be believed in, when we should be living it as a birthday party where all the goodies we are awarded are presents for being born. Let children ask all the questions they want – *the one thing they must never question is their worthiness to be loved.* With all the minutiae that bombards us on a daily basis, at least we can be proud of being at a point in our development as a species where we've got that momentous truth straight.

If love is unconditional, what does that say about heaven? Must it not be unconditional too? If love is heaven and heaven love, it follows the bliss of heaven is as *unconditional* as love. Indeed, if a baby is welcomed with a love that asks nothing for itself, it is safe to say he is in heaven – and heaven is his to enjoy by virtue of his birthright of unconditional love. As it is now for us in our time of dawning consciousness, so it was for us at the dawn of time. *Virtually every religion and mythology speaks of a paradise in the infancy of the race.* Inasmuch as it existed from the get-go, one can conclude no human effort was required to build or inhabit it. Insofar as it served as the *de facto* cradle of man, it is self-evident no arduous or long-drawn-out pilgrimage was required to reach it.

Here's the remarkable thing about Paradise – and even the Bible cannot hide it! *Adam and Eve woke up one morning in Paradise – and damn if they'd done a single thing to deserve it!* They were masters of Eden without a clue as to what they did to get there! They didn't put in the time or work their heads off. They didn't pay for their sins or wade through sloughs of despond or whack away at forests of red tape to fight their way there. Oh no, it was unbelievably simple. *All they had to do to show they had the right stuff was for their dust to be molded into human form.* All they had to do was open their eyes – and there it was! The Garden of Eden in all its glory, waiting for them as if it were the most natural thing on earth! *Far from having to prove how good they were to the world, it is as if the burden was on the world to prove how good it was to them*! Talk about making a great first impression! What does it say about the human race that the first thing that it sees, when it sees the light, is a *world* so committed to proving it is perfect for human beings that it snaps to attention as a paradisiacal thing of beauty and bliss?

In any case, the challenge of paradise is not attaining it but succeeding in not losing it! That is the ever-present threat! Our problem isn't earning our precious inheritance but piddling it all away. It is treating the Holy Grail bequeathed to us as a chamber pot. The truth of paradise makes everything we are taught about success and progress a lie. We don't start at the bottom and work our way up. Our prerogative is to start at the top of the world and take care of one another to ensure we never fall from that beatific height. *Our highest priority is not to further the ascent of man. It is to take immediate steps to reverse the fall of man.* We don't have to be wise or good to deserve paradise – we are naturally that. We just have to be **not** crazy or self-destructive to throw it away!

The professed danger of the participation trophy is as mythical as the danger of handing out an unconditional basic income to everybody. It is not about sapping the work ethic or disincentivizing performance. It is about debunking the rat race for monetary rewards as the only motivator for becoming a functional, helpful member of society. If heaven can be had right out the box, without any preparation or protracted struggle, *being* good is in our nature…as is *doing* good…as is *feeling* good by *doing* good. The challenge is not to be good enough for the good life. It is to live off the superabundance of goodness we bring into the world. As a species with the run of this garden planet, we don't need to win a piece of the universe more fit for princes and kings.[1] We desperately need not to lose it.

---

[1] "Song of the Mira", Allister MacGillivray (1973). Performed by Celtic Thunder

32. WHO NEEDS LEGENDARY CAMELOTS WHEN WE HAVE THE REAL THING

Right out of the gate, we are left with a couple of thoroughly subversive concepts. Paradise or heaven is not a destination for the human race – *it is our point of departure. Forget about going to heaven when we die. We are meant to go to heaven the moment we are born.* The pearly gates open up when the cervix dilates wide enough to let us pass. And since we don't have to pay any dues or jump through probationary hoops to be admitted to this abode of bliss, it must signify we are as perfect as can be from the get-go.

What if we kept this insight front, left and center as we went about conducting the business of society? What if we honored heaven as the birthright of each and every human? What if we made every effort to keep infants in a state of blissful intimacy with their caregivers, instead of taking every opportunity to rudely kick them out of paradise as soon as we can? Acting on the realization that our primal nature is good, what if we gave up trying to improve on it by forcing our children to stop speaking the indigenous language of feelings and learn instead the broken English of self-control and self-denial?

For starters, children would grow up to be a joy and delight. They would radiate the self-confidence that comes of being grounded in their humanity. Their emotional intelligence would not be stunted or blunted. An unerring, enlightened sense of right and wrong would shine through. Overnight children would go from annoying creatures who have to be made to cry to wise beautiful enchanting creatures who bring tears to our eyes.

Beyond that beckons the most earth-shattering discovery

of all. *Our primal experience of heaven as infants and children is the blueprint for creating heaven on earth!* Here we are brainwashed to think that when we leave our homes every morning for our jobs as diplomats or generals, think-tank analysts or bankers, magistrates or guardians of the public trust, we are doing important work. And yet the main reason all these vocations exist in the first place is because we routinely botch the most world-changing vocation of all – parenting. Too many professions exist to pick up the pieces because of our heartbreaking performance as mothers and fathers. They exist in a *janitorial* capacity to clean up the toxic messes left behind by our failure to handle our young with the care and tenderness they deserve. The truth is, much of what we call the fabric of law and order is a straitjacket designed for all those crazy things that many of us wind up doing for love.

Imagine building on the intimate rapport that is the source of infant bliss. The mother-child union, not the United Nations, would be hailed as the bulwark of world peace and solidarity. Imagine if we rejoiced in the smiles of parents and laughter of children. The ranks of the military brass would be as sounding brass and tinkling cymbals. Imagine if the impulse to be contributing members of the family of man could be banked on by honoring our instincts to help out as children. Money would go back from being a slave master to a medium of exchange! What need for regulatory agencies or watchdog commissions if we are raised with the emotional capital to do right by our fellow man by caregivers who did right by our needs for nurture and affection? What if all it takes to lift our brothers and sisters out of misery is being picked up when we cried as children? The closeness and warmth we enjoy as children grows our brain cells and boosts our EQ. Harsh, cold, violent forms of parenting leave us with a defective EQ that is the surest predictor of crime and war. A childhood of emotional bonding and security is not only the foundation of human existence – it is the only real foundation for human *coexistence.*

All we have to do is to build on this heavenly foundation and make sure it is perpetuated over time, without being disrupted or interrupted in any way. It could be said this is what we are put on this earth to do. *Heaven lies about is in our infancy*, sings the poet. But why only in our infancy? Isn't it our job as parents to take the experience of heaven that lies about us in infancy and grow it until it lies about us in our *youth*? Isn't it our job as a society to take this heaven that lies about in our youth and enlarge it until it lies all about in us in our *middle age*? And isn't it our job as a civilization to take this heaven that lies about in our middle age and expand it till it lies about in our *old age.* And when we have run with this experience of heaven until it lies all about us at every stage of our lives, that'll be the day we can sit back and say we have heaven on earth!

There is a reason we are not there already. We act like this heaven is out of reach, when it is embedded in our DNA. Just to make sure we get the message loud and clear, nature goes out of her way to rearrange her schedule to make extra time for a love dispensation. *Heaven forbid we should be left with the suspicion that love is a fly-by-night proposition, here today, gone tomorrow – as it is for the young of every other species.* A baby turtle hatches and find its way to the ocean right away. A newborn giraffe calf clambers upright and is walking around in a matter of hours. A human baby can't even lift its head without help for two months. It takes six months for a baby to sit up, nine months to stand, and a year to take its first steps! The cardinal sin of parenting is to rush children into self-sufficiency before they are ready. Parents apply to their young the shortened time span allotted for a foal or baby chimp to be independent, when in all fairness children should have the extended length of time that is appropriate for, well – a human being. And it may well be that our young fail to live up to their human potential – and eventually attract the label of animals and reptiles – because the length of time giving to nurturing them falls more on the prehuman end of the spectrum than the human one.

When it comes to babying our young, humans are designed to take their own sweet time. This allows for big brains and complex thinking to develop – and more importantly, for big hearts and nuanced feelings to burgeon. Other species *live* longer than man. None is genetically programmed to *love* longer. Of all the instances where man butts heads with nature, none is more detrimental than putting his young on the fast track to independence when nature has clearly geared their growth to the slow, easy track. Given our unequaled allotment of time to meet dependency needs, nature is evidently in no hurry for our children to grow up. And neither should we be. As unstinting as nature is with the superabundance of love that is our birthright, you can be sure she never intended us to short-change our offspring of it. That is exactly what separates us from reptiles and beasts of the field – the *length of time it takes to lay the foundations of love deep in our young and carve out a fully dimensioned capacity to experience it.* Not just our personal stability, well-being and peace of mind but the whole of society's depends on it!

Truth, justice, beauty, peace, goodness – these aristocratic values flower and come to fruition in the expansive light and air and soil of love. *Legendary Camelots just don't make the cut.* Judging by the extraordinary length of time our species has to cherish and coddle, play with and protect our young, nature patently wants us to have the real thing.

33. THE MODEL OF HELL – THE
BASIS OF THE POWER PRINCIPLE

*L*ove as heaven/heaven as love is the norm for organizing society. How on earth was society ever organized
around a power principle? An adult towers over a baby like
a giant, but the giant's strength is at the baby's service. A
parent wields the power of a god, but it is a providential
power that protects the child from things that go bump in
the night and provides a safe haven for a child to laugh and
sing and dance his way into maturity.

When this love structure is disrupted by cosmic violence, the earth-shattering blows that fall from heaven are
interpreted as falling from the hand of supernatural entities. The heavens that shone with the light of providence
now smolder with the flames of perdition. Where the nurture and warmth that emanated from heaven attested to
man's *goodness*, the wrath and chaos that crashes down
from on high bear witness to man's *wickedness.*

When stars run amok and turn the sky into a lake of
burning fire, and the earth into a bottomless pit of smoking darkness, *the heavens become a model of hell.* A new
principle worms its way into the human dimension – *punishment.* Conceived of as a wrathful judgment, a vengeful doom, it never did belong to the human dimension. It
belongs to a celestial realm that wreaked havoc on earth.
The Lord who says *vengeance is mine* is laying title to his
rightful property. Man committed a promethean crime by
stealing the fire of vengeance from heaven – and it would
be better for the good of all if it is promptly handed back.
Humanity should stick with what it's best at – radiating

the light of love – and leave the fire of vengeance to the Oort Cloud from which it emanates in comet form.

As a way of righting wrongs, retribution is the radical game-changer in our moral orientation. *Heaven as love/ love as heaven* marks a normative order of debt. The debt of love must be paid to raise a whole, happy species. The heavens as an infernal abyss, an archetypal hell to which man is exiled by Judgment Day, usher in a new dispensation of debt – debt owed to gods for violating their laws. The enormity of this debt is such that it is first paid by a virtual holocaust of the human race, and subsequently, by an endless stream of burnt offerings, blood sacrifices, mortifying penances and death penalties.

The infernal model of punishment now becomes the basis of organizing society. In happier times, society mirrored a template of heaven as a providential source of nurture and warmth, designed to fulfill the debt owed the children of men. Now it mirrors a template of heaven as a hellish place from which punishment is meted out to enforce payment of man's debt to the gods, and by extension, to a society ruled by wannabe gods. In effect, man is intuited to be good no more. The biological debt of love owed children is no longer the basis for safeguarding the integral order of humanity. Man is demonized as a fallen creature, a sinful, evil being – and the theological debt of life, liberty and happiness he allegedly owes the gods is the legitimate basis for maintaining the social order.

You'd think a debt that serves as a warrant for genocide is of a staggering order. Who'd ever believe it is a *phantom* debt! Much less that this *phantom* debt is the sacred foundation of a civilization that serves as a glorified collection arm for *phantom* beings! Just because this debt is imaginary doesn't mean it's harmless. As king and priest become ruthless collectors of fictitious debts owed to gods, what becomes of the all too real debt owed humanity? It gets short shrift – with disastrous consequences. We hardly allow for the possibility that if man is guilty of violating the laws of gods, it is because the laws of humanity cannot be obeyed by those who are violently cheated out of their due of love.

Our *biological* debt of love is a natural thing. The expectation of having this debt paid is in our genes, and engenders a felicitous fulfillment that leads to a society based on cooperation and comity. Our *theological* debt is another matter altogether. It does not come organically to man. The infernal destruction from comets, meteors, supernova, or whatever, has to be conceptualized as the wrath of a class of genocidal maniacs known as gods. This divine wrath places man under a form of indebtedness known as *sin*. Unlike the payment of the debt of love, which is a source of joy and fulfillment that man needs and cannot flourish without, the payment of the debt of sin is a source of fear and shame that causes man to fall under a death sentence. And under the terms of the *modus vivendi* worked out by religion, man incorporates the burning wrath of gods and turns it against his own flesh and blood. He arms himself with the firepower of gods to make burnt offerings of his kind or burn them at the stake. He condemns them to the flames of hell in the hereafter or to the firestorms here on earth of the hell that is war.

Once the cataclysmic violence of heaven is framed as the judgment of the gods, and man emulates it by using death and suffering to pass judgment on his fellow man, it alters the whole structure of society. Compassion and kindness require no mechanism of force or violence, as they answer needs that spring from deep in our biology. Divinely inspired punitive judgments require the institutionalization of force and violence, since they come from outer space and derive their mandate from an alien construct known as theology. *The infernal model of punishment is thus the origin of the power principle.* Punishment implies more than superior moral authority. It implies enforcement of that authority through superior size, strength, cunning or force. The only way to punish is through an edge of superiority. Parents who pay their debt of love are not threatened when a child shoots up to be their equal. The same cannot be said of parents who wield a rod of wrath to collect a debt. An adult who makes the child pay a fictitious debt can get away with punishing a weak and small child. Once the child grows into adulthood, the disparity in size and strength necessary for the administration of punishment disappears!

Institutional guardians are not invested in paying the biological debt of love owed to mankind. They are invested in setting themselves up as punitive gods, sitting in judgment on mankind. Hence they must take the natural disparity between adult and child and replicate it through an array of technological or tactical artifices in order to *perpetuate* it between themselves and society at large. If an adult looms as a god to a child, guardians must loom as demigods to a community of adults. Only by embodying the strength or force (kratos) of gods are they in a position to enforce the *punitive principle.*

Discharging a biological debt through nurture and care promotes intimacy and camaraderie. Collecting a theological debt through vengeance and wrath calls for dominance and force. The infernal model of punishment begets a will to power that triggers endless games of power-seeking. The prerogative of passing punitive judgments on society is partially upheld by assuming the status of a god – it is absolutely upheld by acting as a type of supreme God who overpowers all others. As a mania to imitate gods in the wake of Doomsday makes *vengeance* and *retribution* all the rage, the ritual of jockeying for power in order to *judge* and *punish* becomes the order of the day. This means that as the natural disparity between parent and child is forever narrowing to zero, the artificial disparity between leader and society is destined to be perpetually widening to infinity!

In short, once the punitive principle gives birth to the power principle, the race for power is on in a never-ending quest for dominance, a spiraling struggle for superior force. The history of civilization becomes a dreary, interminable round of oneupmanship to see who is going to enforce the collection of punitive debts on society's behalf – exactly *who* is going to make *who* pay as the closest facsimile to a god of vengeance. Of course, these wannabe gods merely follow the lead of doomsday gods in acting as though children are indebted to parents and guardians, instead of the reverse, and go about collecting this debt through institutional hells like prison and war. The upshot is not a power structure that can bring itself to use the wand of *love*, as per the dictates of *biology*, to manifest the better *angels* of our nature. It is a power structure whose knee-jerk reaction is to wield the rod of *wrath*, as per the dictates of *theology*, to conjure up the potential *devil* in mankind.

Kings are seen as gods but what is unseen is that as gods

their domain is *heaven* not earth – and not the heaven from which sunbeams and raindrops fall to make earth a garden planet, but the heaven from which fire and brimstone fall to make it a cindery waste. Rulers who arrogate the judgment of gods are on a slippery slope to recapitulate the madness and mayhem of a Last Judgment. The guardians of religion could not very well settle their debt to the gods without modeling their destructive behavior in acts that verged on madness, depravity, and crime.[1] What wonder that the secular guardians of law and order who make mankind pay for its crimes and transgressions are, in fact, the ones guilty of turning history into a dossier of crimes and transgressions against humanity.

---

[1] Mircea Eliade, **The Sacred And the Profane** [p. 104]

34. THE QUEST FOR FULL-SPECTRUM DOMINANCE – OR IS IT *DOOMINANCE?*

M an can only mete out *doom* through a quest for *dominance.* Power is no stand-alone principle, surviving on its merits. Its justification from the start has been as a means to administer punishment – a way to *correct* man by *erasing* his humanity. The raison de étre of the power structure is to punish evil. It restores the balance of order by collecting man's debt to society. It forces man to pay for his crimes with his freedom, dignity and life. God as a devouring fire is incorporated into sacred rituals by immolating people on hilltop altars or in autos-da-fé in the town square. The sword of comets is incorporated into public policy and military strategy by putting people to the sword. A Day of Wrath is perpetuated down the ages by deploying engines of annihilation like tomahawks, tasers, trebuchets, tanks, guided missiles, gunboats, gallows, and gas chambers. Relatives of murdered victims await their day in court to make it a Day of Vengeance.

The conquest of space is the climax of this story. The first rockets were not built for glamorous adventure or science exploration. They were built to rain death and destruction on Londoners. The U.S. was quick to seize on the advantage of V2 rockets. As soon as the war ended, pioneers of the German rocket program were smuggled in to work on the U.S. space program. The Joint Chiefs knew what they doing. They had a doomsday weapon but no good delivery mechanism to unleash it on the enemy. Thus rockets evolved into ballistic missiles. And the expertise of Nazis who launched rockets to destroy London was tapped to create intercontinental ballistic missiles to destroy Moscow.

If the desire to resemble Supernatural Beings tormented man from the beginning of history,[1] nothing promises to ease that torment more than a space program. Shamanic myths of magical flights to heaven...a sacred quest for absolute freedom by bursting earth-bound limitations...this is the *essential nostalgia* at the heart of religion. Thanks to advances in astronautics, the sky is the limit for indulging it. This race for transcendent ascendancy is powered by a will to bring the world under the control of a new breed of wannabe gods. Its rationale as always is to punish those who transgress the party line, or offend the ruling elite, by wielding a rod more befitting the hands of gods than of men – death stars and orbiting lasers and space-based nukes to wipe out continents in a flash.

Of course that doesn't go over well in a democratic age presumed to be enlightened enough to have junked the cult of heaven and the archaic imitation of gods. So military conquest is not shown for what it is but sexed up as the pursuit of freedom and security. The conquest of space is not shown for what it is but cleaned up and presented as a noble ambition to explore the universe and expand the bounds of knowledge. But let's not harbor any illusions. If the National Aeronautics and Space Administration works hand in glove with the Department of Defense, it is because NASA is DOD's sock puppet. The DOD chooses the direction of the space program. Lunar rocket and modules are built by the same firms contracted by DOD to build bombers and missiles. The men of Apollo who were

---

[1] Mircea Eliade, **Rites And Symbols of Initiation** [p. 101]

supposed to come in peace are military-trained personnel. *The olive branch is the ultimate fig leaf.* Space is not a proving ground for heroic adventure and scientific advancement but a theater of war. And as surely as Great Britain cemented its empire through a mastery of the seas, the U.S. cements its empire through a mastery of space.

Nobody realized this better than one-time SS member and rocket scientist Werner von Braun. He was the guiding force behind the development of the U.S. space program. When he and his colleagues dreamed up the idea of a space station, it was not a jumping off point for going to the moon. It was a *space fortress* for achieving world domination. "If you consider that the space station flies over all the populated regions of the earth you see that a nuclear war technology of this kind gives the builders of the space station the most significant tactical and strategic advantages ever known in the history of war."[1]

And thus the same old quest for a tactical and strategic edge, the same old hunger for commanding heights, which drove feudal lords to build stone fortresses on hills, still drives their modern descendants to build high-tech fortresses in space. However much the PR campaign may beg to differ, the space program is not about the conquest of space. It is about the conquest of the world. It is not about developing the capability to put man on the moon. It is about developing the capability to dispossess mankind of the earth.

And it is a safe bet that the U.S. has succeeded much more effectively in the latter goal than in the former. The

---

[1] *One Small Step – the Great Moon Hoax and the Race to Dominate Earth From Space,* Gerhard Wisnewski [p. 64-65]

bungled history of the Apollo program – the mind-boggling lack of testing, the embarrassing record of rockets that kept exploding and lunar modules that kept malfunctioning – proves that NASA wasn't really serious about sending man to the moon and bringing him safely back – except as a massive publicity stunt to wow the world. If the moon was the real objective, the prudent and methodical thing to have done is build a space station first before tackling the logistics of getting to another planet. Instead, NASA did everything backwards. It rushed to complete a manned lunar mission and then set about building the space station as an afterthought – which is like setting out to scale Mount Everest first and then returning from the summit to establish base camp!

Once NASA's military orientation becomes evident, the idea of man going to the moon doesn't seem like Holy Writ any more. Ask yourself what is the first casualty of war to suss out why NASA has become an acronym for *Never A Straight Answer*. Were NASA about scientific exploration, the chances of being lied to would be minimal. Were it about an ascendancy of power, its level of lying would rise to that of a politician. But as its thrust into space is geared to war, its lying quotient shoots up to astronomic levels.

In any case, since when did the real world ever intrude on megalomaniac dreams of world domination? When did reality ever rain on the charade of playing god? From the outset the enterprise to awe and cow is capitalized by the cult of the Big Lie. A king as beneficent dispenser of nature's gifts no more exists than Santa Claus. In order to establish sovereignty in the ancient world, a being created by the ordinary fertilization of an ovum – and locked in the frailties of parental genes – is transfigured into an emanation descending from the clouds or the sun.[1] And in order to establish mastery in the modern world, a program hatched between a mad Nazi scientist like Wernher von Braun and a Hollywood dream manufacturer like Walt Disney[2] – and locked into the failings of an underdeveloped 1960's technology – is transfigured into an epiphany rising on billowing clouds of fire from the earth to the moon. Honorific epithets and grandiose titles with a tenuous grip on reality can nonetheless burnish a god-king's image and bolster his divine pretensions. One day the myth of Apollo may lie shattered like the hulking remains of Ozymandias, lost in the boundless wasteland of space, but it will have long ago spectacularly served its propaganda mission of clinching the argument for American superiority.

Taking liberties with the truth is to be expected of a mindset that insists on breaking free from the bounds of earthly reality. As long as a soaring pyramid does an unforgettable job of showcasing Pharaoh's glorious journey into the next world, what does it matter if its mission is a monumental dud? As long as millions are convinced that man can defy his earthly limitations to survive, like Avicenna's Floating Man, in the vacuum of the moon, what does it matter if Apollo is shot on a sound stage in the Nevada desert?

The he-man myth of self-reliance has always been about jumping the gun and going off half-cocked by pushing the

---

[1] Barrows Dunham, **Man Against Myth** [p. 218]. Little, Brown and Company, 1947

[2] **One Small Step, The Great Moon Hoax And the Race to Dominate Earth**, Gerhard Wisnewski [p. 71]

children of men to survive outside the protective mantle of mother earth before they are ready or able. It has always been about wishful thinking and the bragging rights that go with it. The propaganda of myths and illusions, fables and delusions, is the fiat currency of imperial godhead and global domination.

One picture of the American flag on the moon is worth a million words painstakingly spent debunking it. Is the White House or Pentagon featured on the cover of a book that makes the case for U.S. geostrategic primacy and world domination?[1] Is it troops raising the flag? A bomber fleet? No! It is a U.S. astronaut on the moon saluting his flag with a lunar module beside him. There you have it. The space-age equivalent of a stele raised in honor of the Babylonian emperor and carved with myths to glorify his exploits.

---

[1] Gerhard Wisnewski, *One Small Step – The Great Moon Hoax* [pp. 73-74]

35. THE LOW-TECH SPACE PROGRAM

Earth is the cradle of humanity, but one cannot live in a cradle forever.[1] So says the father of astronautics. But what says mother nature? *Humanity has to linger in the cradle longer than any other species.* Who do we go with then, father of astronautics or mother nature? Is it too much to expect the children of man to be tied to the apron strings of mother earth until they can sally forth to explore the world without being lost in space? Can we at least live in our earthly cradle until we learn to crawl through a fusillade of galactic rays without coming to harm? Or walk on the moon without having our spacesuits punctured by the relentless hail of micrometeorites that pulverizes its surface to fine dust?

The fittest response to a U.S. President's rash promise to put a man on the moon before the decade's end would have been the Elizabethan proverb – *look before you leap.* Runaways gamble on the cold mean streets being preferable to an abuse-wracked home. The 60s were a time when the flight to a bleak, forbidding planet seemed a good way to escape a country wracked by wars, assassinations, race riots and foreign-policy debacles.

The headlong rush for parts extraterrestrial is rooted in a deep-seated amnesia. It is long forgotten that before man went into space, space came to man. Asteroids tumbled to earth. Comets ripped the cocoon of the atmosphere. Effluvia of supernova engulfed the earth in radiation. Solar flares and cosmic galactic rays slammed into the ground,

---

[1] Konstantin Tsiolkovsky

sending out high-velocity microscopic shrapnel, lethal to animals, plants and people.[1]

The hostile penetration of space into the sphere of earthly existence meant that humans were no longer earthlings per se, but *de facto* spacemen expected to survive the rigors and ravages of space. Survivors of extinction-level events lived in terror of their recurrence. They might never have guessed their descendants would forget the horror of it all, many millennia later, and embrace the prospect of throwing off earth's protective mantle and voluntarily returning to the place from whence their doomsday nemesis came.

Except for one thing. The horror of being denuded of the sanctuary of the terrestrial mother may have felt like the end of the world for mankind. But it wasn't the end of it for future generations. For it led directly to the horror of being denuded of the sanctuary of the individual mother and plunged into a void almost as desolate as space. NASA can be forgiven for thinking man can be wrenched from the arms of mother earth and sent hurtling into the immensity of space. After all, didn't the first heroes survive that way? Didn't our ancestors make it through a pit of desolation when earth's protective folds were ripped away and man stood naked to the dire assaults of the cosmos? Our existence as a species is proof we survived the worst that space dished out by way of lethal bombardments and lived to tell about it. No wonder we lost our appreciation for how good we have it on our host planet. Nature goes out of her way to provide every comfort on earth and we cynically and

---

[1] *The Cycle of Cosmic Catastrophes*, Firestone, West and Warwick-Smith [p. 135]

cavalierly think we can blow it off and do just fine without it.

The toxic fruit of that thinking is the he-man ideology bred into culture heroes. *I can live without my mother's protective embrace and brave a terrestrial wilderness devoid of the milk of human kindness, the honeyed sweetness of life.* The space age carries it a bridge too far. It deepens the contempt for the maternal universe to cosmic dimensions. *I can survive without mother earth's swaddling bands, forsaking my cushy place on this garden planet to brave an extraterrestrial wilderness – the killing fields of cosmic galactic rays, solar flares, meteoroids, and radiation worse than inside a nuclear reactor.*

The fallout from Doomsday made a proto-astronaut out of man. Religious man is hurled into a lethal space where the fundamentals of existence are solitude, danger and hostility of the surrounding world.[1] In turn, the desolate space of religion – bristling with malice and menace – became the cradle of the 'religion of space.'[2] And the mediating agency is the culture hero – abandoned to the dangers of the wild and left to survive in a hostile environment. Long before the sky was the limit, the ideology of culture heroism was hard at work pushing man to test his limits and prove his survival skills. The bravado of brinkmanship is the proving ground of civilization. NASA pushed it to a breaking point, but being smart enough to know that flesh and blood wouldn't pass the test, did the practical thing

---

[1] Mircea Eliade, **Shamanism** [p. 27], Bollingen Series LXXVI, Princeton University Press

[2] G. Wisnewski, **One Small Step – The Great Moon Hoax and the Race to Dominate the Earth** [p. 80]

and faked it. But it wouldn't have even dared to think of going to such insupportable extremes without a long history of culture heroism setting up shop as a divine authority over nature, and upending the conditions of life, birth, growth, maturation.

Nature decrees that man is a product of love. Two people join in ecstatic union and plant the seed in a matrix that is exquisitely congenial to the creation of life. After receiving the stream of nourishment needed to grow to fruition, a nascent being emerges from this protective space and continues to receive the flow of nurturance needed to grow into a fully realized human being. Postapocalyptic man will have none of that. From the tribal to the civilized stage, man believes he can do better. The ideal matrix must be harsh, cruel, inimical to life. The ideal womb is a crucible of dire privation where life survives by aborting many of the functions that make it quintessentially human. When this creature emerges at last, it has lost much of the softness and sweetness, tenderness and vulnerability that defines a human being. Only now is this monster hailed a 'man.'

The completion of our journey through a uterine cocoon is an occasion of great joy. Its anniversary is celebrated as long as we live. The passage through a gantlet of horrors that is supposedly *the making of us* marks our emergence as *real* men. *How many roads must a man walk down before they call him a man?* In real life, just one. He must safely pass the birth canal and be raised in a supportive milieu. In the realm of myth, he has to walk over burning coals or blaze a trail through feral jungles or frozen wastes. In the realm of legend, a man is not a man until he takes his life in hands to undergo a baptism of fire in a milieu abusive in the extreme, a world out to annihilate him. Forgetting he is created in a womb and fulfills his potential in a loving facsimile of it, he labors under the delusion that he proves his manhood by testing himself against nature at her harshest and most inhospitable. Men who brave the Arctic wastes to reach the North Pole, or the icy slopes of Everest to reach the summit, are awarded medals and knighthoods. Men who brave radiation blasts in the vacuum of space, and are televised hopping through a lunar wasteland, are welcomed as conquering heroes and feted with ticker tape parades.

Is the glorification of the he-man at the bottom of an inflated astronaut cult? If we weren't in the habit as a culture of deriding as unmanly an affinity to the maternal ground of our being, would we be so devil-may-care as a civilization to think we can just shuck off the swaddling bands of mother earth and venture into a universe all wrong for life?

In one sense, it is an epochal breakthrough – in another, not so much! Progress is meant to make life easier, more comfortable. Space travel flies in the face of progress to make a grim mockery of it. It hurls us from a planetary Eden where we breathe easy as zephyrs fan our cheek, sunlight warms our body, rain fructifies our soil, to an abysmal pit that is dead set against life, a cannibalistic void where breathing requires no end of trouble and artifice, temperature extremes boil the blood and crack the skin, and seas of radiation doom us to instant death. How does enduring this celestial hell-hole forge a hero? The answer takes us back thousands of years. Our ancestors lacked the know-how to defy gravity and send men into space. They had recourse to something more primal.

They whisked babies from the maternal ground of being to a wasteland that may as well have been lunar as earthly in its desolation. Whether this made them heroes is an open question, but it unquestionably left them without the mothering to be fully human.

The future is now in the hands of astronauts that *carry the fire*. They are bundled into capsules and shot into a void in thunderous rockets that guzzle millions of tons of liquid fuel, run through billions of dollars like water and require a vast coordination of interdisciplinary talent to assemble, program and launch. Having no such resources, our ancestors did well enough. They didn't have to encase man in a Teflon-coated spacesuit. They could wrap a newborn in animal skins and expose him on a mountainside. They didn't need rocket science to blast a man into orbit. The sheer intensity of their paranoia was sufficient to carry an infant into the forest and abandon him. They didn't need a space capsule to carry a man into the celestial depths. They had an ark of bulrushes to set a baby adrift on the current. For all that it was done on the cheap, and in the lowest-tech way, they managed to do something much more momentous than bursting the bonds that bind us to our mother earth. *They shattered the bonds that bind us to our earthly mother.*

Astronauts return after defying gravity in low-earth orbit and resume their lives with little or no ill effects. What of those who defy the bonds of love? By the time they return to society, they have missed out on critical phases of human development. Adapting to a dehumanized state, their biochemistry is altered in subtle and not so subtle ways.

The first thing to notice is the light goes out of their eyes. Without eye-to-eye soul-gazing interactions we lose touch with who we are. Our ability to know ourselves is benighted. We go *dark*. Loss of tactile contact leaves us without the mammalian warmth and intimacy necessary for maturation. Finding it difficult to form close, trusting attachments, we project an aura that is distant, *remote*. Failure to meet primal needs leaves us unfulfilled, *empty*. Without the other by our side, we cannot feel our feelings and put them into cold storage, growing *cold*. Unable to source anger and dissolve it into grief, we act out the toxic rage that builds up in us by becoming aggressive, violent, *hostile*.

As it happens, this constellation of qualities – *cold, dark, remote, empty, hostile* –characterizes space once we leave the protective embrace of mother's earth atmosphere and magnetosphere. Insofar as these qualities characterize those ripped from the maternal universe and left to fend for themselves in the wild, they can be said to embody a *deep-space psychology*. It turns out our ancestors didn't require towering gantries and rocket engines to stage the most epochal breakthrough in history. They didn't have to engineer the hardware of space flight to put man into deep space. Not when they had something more earth-shattering at their disposal – *a way of putting deep space into man*.

V. BREAKING OUT OF THE MILGRAM PARADIGM

Since wars begin in the minds of men, it is in the midst of men that the defense of peace must be constructed.

— Preamble to UNESCO Constitution

The call to abandon illusions about our condition is a call to abandon the condition that requires illusions.

— Karl Marx, **Early Writings** (1843)

36. ASSIMILATED TO THE GODS
THROUGH CRUCIBLES OF WRATH

The movement of stars and planets is like the orderly flow of traffic. Motorists obey traffic signs and signals, show courtesy to others, drive defensively, reach destinations safely. Every once in a while there is a lapse of attention, a case of driving under the influence. Vehicles veer off lanes, run stop signs, fail to brake in time – and there is a smash-up with attendant loss of life.

This is what happened to our cosmos. Stars exploded, comets drifted into earth's orbit, asteroids flew off the guardrails. The planet suffered a smash-up that decimated our species. Under normal circumstances, it would be ruled an *accidental collision*, with no malicious agency involved. Survivors were in no condition to think clearly. Congested by trauma, and left confused and helpless by such a terrible event, they could have laid down and died. Instead, they valiantly struggled to make sense of their plight by making paranoid leaps of the imagination.

The *first* leap was to imagine that the forces of destruction were intelligent, conscious entities known as deities, not the brute, insentient forces of nature. The *second* leap was to imagine these gods in the grip of a preternatural anger that led them to commit a premeditated act of aggression and malice aforethought. And though their earth-shattering violence fell indiscriminately on many species, the *third* leap was to imagine that it was directed specifically at the *human* species for some terrible iniquity, some egregious lapse from the norms of good behavior.

This triple feat of imagination locked civilization into a tripartite pattern in which humanity has been fundamentally stuck ever since. It set up an imaginary class of beings, possessed of quasi-divine powers, as guardians of human law and order. It enshrined violence as an acceptable method of teaching man a lesson. And it instituted an ideal of justice which equated the death and destruction visited on the human race as an expression of *doom* or righteous *judgment*.

Ages have gone by and these ruling principles remain the cornerstone of the power structure. *Guardians of law and order still project a superhuman power or authority. Violence is still the educational tool of choice by which we teach a lesson. And the enforced desolation and destruction of human life is still the method by which corrective justice is meted out to wrongdoers.* And we have so far forgotten that these principles originated in the hyperactive imagination of doomsday survivors that we have institutionalized them as the height of real-world pragmatism and derisively dismiss as starry-eyed dreamers those who point out there is a more sensible way!

In effect, the religion of power derives from the power of religion. Without religion, the catastrophes that befell man would be the violence of the cosmos playing out according to laws of astrophysics. *The optics of religion saw these cosmic accidents as acts of retribution!* The genocidal violence of gods is no longer unleashed on an innocent species, going about its business. Religion had to go and legitimatize it by saddling man with a non-existent burden of evil. Thus *imaginary* beings identified as the perpetrators of this extinction-level event are deemed righteous and just. And its *all-too-real* innocent victims are damned with an *imaginary* guilt and culpability!

If some malevolent demiurge had sat down and planned

this out, it couldn't have led to a more disastrous outcome. Gods guilty of genocidal vengeance are worthy of worship and adoration. Their innocent victims are deserving of death and destruction! The lie religion perpetrated is not the *Big* Lie. It is the *Cosmic* Lie. *Deify* the brute, insentient forces of destruction as righteous and good. *Demonize* a conscious, intelligent species as wicked and evil! Religion didn't just stand justice on its head. It did something infinitely worse. It presided over the shotgun wedding of judgment and apocalyptic doom – the ideal of justice and ignominy of punishment.

We have never lived down this virulent legacy. As a rule love fulfills our nature, so much so that the very lovability of our nature is kept alive through love. If we act in hateful ways, it is a sign that our need for love is not fully met. Imputing world destruction to wrathful gods radically changes that perspective. The lapse from standards of humanity is viewed not as a distress signal but as a sign that *man is asking for it* – the *it* in this case being the infliction of further distress! Instead of a poignant cry to be *cherished,* it is now a provocative demand to be *punished.* The bottom had to fall out of our morale as a species for such a sadistic shift in morality to occur.

Man evolved with a natural genius for the science of human relationships. We knew how to keep each other wholesome and whole with the warmth of physical closeness and radiance of kindness and affection. Our genuineness is inextricably bound up with the experience of love. But then another standard came into play – a standard of purity, from the Latin *purus*, which derives from the Greek *pyr*, fire, as fire was originally used to cleanse. And for that we have to thank the cosmic holocausts that made a *pyre* of the earth and thereby redefined the imperative to be *pure*. Purity became a religious conceit of divine origin, a sterile, hollow ideal that can only be achieved by passing the world, the flesh and the devil through the fire in rites of purification. It is not enough to keep our sanity through love. Humans have to be sanitized through ordeals by fire. The Latin *castus*, meaning *pure, chaste*, is the taproot of words that denote a whole calculus of punishment of escalating degrees of severity – from *chasten* to *chastise* to *castigate*. We no longer cherish an organic innocence with easygoing interactions of mirth and joy. We set about banishing a *mythical* evil in quest of a *mystical* purity through the methodical infliction of pain and humiliation. And the ultimate grade of punishment is to burn in a *fire* that never goes out.

Divine fires of wrath purify man in a crucible of punishment. The real significance of the discovery of fire lay in a *spiritual*, not physical, mastery over nature. "Fire purifies all things that are brought near it, releasing them from the bonds of matter…making them meet for communion with the gods."[1] A celestial holocaust burned away man's flesh-and-blood connections to align him with gods. Man ever after passes through baptisms of fire to slough off his physical bonds like so much dross. It is not merely that kings and commoners aspire to divinity through death by fire. Commoners cannot even be assimilated into a divine order of kings without being consumed by fires of wrath, *transmuting their humanity by a fit of aggressive and terror-striking*

---

[1] James George Frazer, ***The New Golden Bough*** [p. 368] Theodor H. Gastor

fury. "The hero is the man in fury, possessed by his own own tumultuous and burning energy."[1]

As a practical matter, man could not be expected to put out the flames of a cosmic holocaust. But he *can* definitely put out the fires of wrath in his soul. These fires of wrath are an abiding source of terror. "The wrath and heat induced by a violent and excessive access of sacred power are feared by the majority of mankind."[2] The *shanti* of peace comes of extinguishing this wrath. And that is well within the therapeutic power of each and every one of us to do. But *not* as long as we are being institutionalized to think that the way to bring peace and tranquility to the world is by making it reverberate with the bangs and booms of an apocalyptic holocaust.

---

[1] Mircea Eliade, ***Rites And Symbols of Initiation*** [p. 84]

[2] Ibid. [p. 86]

37. IMAGINEERING EVIL INTO EXISTENCE

The punitive principle ushered in by religion revamped the social order. Punishment has two prerequisites. *Physical dominance. And the high moral ground.* Both were the monopoly of gods. Conversely, the destructive wrath visited on man cast him in the role of an abused child. With no one one his side, man pulled a Job in the face of an abusive God. Beset by tormentors who lacked the first clue as to how to advocate for his cause or support his case, he split off from his victim self to identify with his abuser. This identification with superhuman bogeys who are guilty of preternatural levels of abuse has terrorized mankind from the dawn of history.[1] Regrettably, it was a package deal. Man could not emulate the gods without embracing their punitive dispensation and embodying the physical dominance and moral superiority that go along with it.

We mindlessly accept the proposition that punishment is about deterring lawbreakers and enforcing the law. Never mind that by punishing we break the most fundamental law of humanity – our unity as a species. Punishment is at its core a triad of hubristic claims: *I am not as you are. I am better than you. And if I faced the exact same set of circumstances as you, I would handle myself better than you.* "When you say I am not as other men, you have lost the two most valuable qualities we have tried to attain – humility and brotherhood."[2] It's not as if we have to attain these

---

[1] Mircea Eliade, **Rites and Symbols of Initiation** [p. 101]

[2] **They Came to Baghdad**. From a collection of Agatha Christie stories, **Spies Among Us** [p. 94]

qualities as they are built into human nature. Punishment is an alien concept to our species that requires us to betray the common touch that makes us all kin. Seduced by the imposing facades of capital houses and court buildings, the bully pulpits from which commanders-in-chief deploy generals and soldiers, and the judicial bench from which judges bark commands at bailiffs and gendarmerie, we lose sight of a simple fact. The *impulse to punish* is inseparable from the *will to power*. The exercise of moral superiority is bound up with physical dominance.

If humans were meant to punish, children would be eternally weak and small, and parents forever big and strong. But the goal of parents to raise children to be equal in size, stature and strength to them undercuts the whole basis of punishment...whereas the goal of institutional guardians to tower over the rest of humanity, by amassing greater and greater power, expands and reinforces the superiority of adults over children. *Even the growth of democracy is powerless to reverse or arrest this trend.* While we flatter ourselves that political leaders are just like us, their high-tech power to control, surveil and destroy mankind has grown exponentially over the last century, to the point of putting the overbearing kings of antiquity to shame. People have an inalienable right to life, liberty and the pursuit of happiness, but the evidence shows that a tiny cabal bent on playing god is obsessed with death, imprisonment and the pursuit of dominance.

In any case, to glom onto punishment as the solution to evil gears the whole penal system to bringing the fantasy of evil to fruition. Interplanetary discharges are the thunderbolts of Thor or Jupiter. The stone or rock crashing down from the sky is the iconic hammer that strikes dead, personified as Death or the Devil in the Teutonic word *Hamar.*[1] As this hammer passes into the mortal hands of authority figures who make it their holy mission to impersonate gods, the mandate to deliver man from evil takes a back seat to repurposing the home into a petri dish for actively growing a culture of evil to justify the wrathful judgments of gods. *Judges cannot be armed with the hammer of justice without propagating a tendency to see evil nailed everywhere.*

The proverbial rod didn't sprout up in a vacuum. *Earthly* parents raised a punitive hand against generations of young in imitation of *heavenly* parents who were believed to do it first. As children are born as innocent as mankind itself was in the beginning, parents wage a campaign of abuse against the young by projecting all manner of evil onto them, imagining them as imps of chaos, spawn of the devil. And nothing is guaranteed to raise children as creatures of darkness more than the traditional practice of beating the daylights out of them. The proverbial rod, in fact, is a black magic wand for turning the young of our species from cuddlebugs into bugbears. The lesson that children take away from punitive discipline is to deaden their feelings and become unfeeling disciplinarians in turn. The impetus to *instruct* children with wrath is a tried and true technique of raising them to be *instruments* of wrath. The violence that beats children into insensibility in the home is the bloody fountainhead for senseless forms of violence in society.

Punishment would have been phased out ages ago were

---

[1] Jacob Grimm, **Teutonic Mythology**, Volume One [pp. 180-182] Dover Publications

it not for a simple snag. *The use of punishment to curb 'evil' in the political arena is heavily dependent on its use to curb 'evil' in the domestic arena.* Money may be the lifeblood of the economy, but the currency of childhood trauma is the proverbial bad blood of the political economy. A vindictive spirit literally wills into existence any evil that the paranoid imagination dreams up. Make conditions in the home hostile to the experience of love, and you can churn out any number of unloving people to satisfy the most vociferous boosters of penal incarceration, capital punishment and military aggression.

The Three Wise Monkeys who see, hear and speak no evil are the antithesis of a mainstream media that eats, sleeps and breathes evil to justify an institutional mandate to channel the gods of wrath. Our guardians are groomed to play doomsday gods by taking the fantasy of evil which triggered a Day of Doom and working around the historic clock to make it come true. We are so used to incorporating the wrath of gods into the human sphere that we have forgotten its delirious origins. The wrath of gods was never a real thing. It was the construct of a feverish imagination. But once imagined, it is necessary to justify it by making our species *evil*. Hence evil is not a real thing either. It is a derivative conceit used by survivors to justify divine wrath. As soon as culture heroes began serving as conduits of divine wrath, in their capacity as warrior-kings, *individuals* or entire *groups* had to be seen as evil to deserve that wrath. By the same token, if man is evil enough to be targeted by divine wrath, what does that say about the gods who see fit to punish man? Why, they must be paragons of *goodness*! So in the same breath that religion made man bear the brunt of guilt for being a creature of chaos and darkness, the gods were absolved of their role in demonizing man and came to be revered as creatures of order and light.

This is the cosmic blindspot of religion. The worship of gods is rooted in the debasement of man. Gods are exalted to the skies by treading man underfoot. Religion gets away with it by a brazen attempt to repeal the laws of physics and enshrine a mystical principle of punishment in its place. Once punishment becomes an immutable law, it takes on a logic and momentum of its own. What sets death stars in motion are not the laws of physical motion or inertia but the idea of *evil* in man. The idea of *evil* is powerful enough to reorder the physical universe. Having once attributed real existence to an idea this powerful, the mind wants to see it alive and can effect this only by personalizing it.[1] And what better way to bring evil to life than by punishing rituals that deaden our sensibilities under the guise of disciplining them? What better way to personify the effect of evil than through depersonalizing regimens and dehumanizing ordeals? The religious conceit that the most powerful forces in the universe are *mad* at man predictably evokes behaviors designed to lend it a veneer of credibility. The disastrous force and energy that slams the rod into their behinds not only expels children from the maternal universe but propels them beyond their intrinsic order to a space which is of the essence of evil – cold, dark, empty, remote, hostile. William Blake summed it up best: *What is now proved was once only imagined.*

Punitive laws and judgments are not a defense against

---

[1] J. Huizinger, *The Waning of the Middle Ages.* George L. Mosse, *Toward The Final Solution* [p. 233]

the evil in man. The evil in man is a defense against the pain of punitive laws and judgments. The ancients had it half-right. Whom the gods would destroy they first make *mad.* God knows there are abusive practices aplenty to drive children mad, and once they grow up *mad,* there is no end to the *bad* things they do. The will to imitate or identify with gods comes with a lot of baggage – vengeance, retribution, punishment. To justify the idea that the mass death of our species resulted from the law of retaliation, not the laws of astrophysics, it is necessary to imagine men are evil, then imagine the evil in men into existence. Since no one is born evil, evil must first be projected on innocent souls, who must then be driven to act out in evil ways in a classic self-fulfilling prophesy. In due course, the sum total of evil in man represents a cumulative attempt to become worthy of the wrath of gods.

Funny, isn't it, all the time, money and effort expended by our legal system on proving the guilt of members of a species who are factually innocent to begin with. If you want to imagine something, imagine this. If a mere fraction of these resources were expended on protecting the innocence of our young from birth, there would be no more guilty people left to feed into the institutional belly of the beast, and it would die of starvation. If you would rather not imagine that, you are still left with this. *If evil is not innate in man, then it follows that all the evildoing painstakingly proved in courts of laws was once just an ill-starred twinkle in the eyes of parents.*

38. EMOTIONAL EMASCULATION

The Book of Genesis opens with an enigmatic link between punishment and the Tree of Knowledge of Good and Evil. What do they have to do with each other? The answer has proved elusive because the focus has generally been on ID'ing the Tree of the Knowledge of Good and Evil. To unriddle the link, the focus has to shift to the structure of punishment itself. Think of the act of punishment and two things spring to mind. Dishing out the punishment and undergoing it. The party dishing it out is supposedly righteous – the party undergoing it is supposedly wicked. One party good, the other evil. Shades of the Tree of the Knowledge of Good and Evil!

Love is a unifying principle. To be loved unconditionally is to be loved as a whole. We are loved not just for *who* we are but for *all* of who we are. We don't take sides, or pick and choose one part to love, another to hate. No part of us is less good or more evil than another. All that changes when punishment comes into the picture. Punishment drives a wedge into our integrity as individuals and our unity as a species. It divides us against ourselves and against one another, teaching us to know one part of our humanity as righteous, another part as reprobate.

At last the confusion clears up. We are not punished for eating of the tree of knowledge of good and evil. We partake of the tree of knowledge of good and evil by participating in the ritual of punishment. *The nature of punishment itself ensnares us in this malignant knowledge.* Man cannot *ascribe* a punitive agenda to gods without succumbing to a fatal knowledge of good and evil. Nor can he *subscribe* to a punitive agenda without splitting asunder the integrity of the human continuum. One part of mankind is elevated to the high moral standing of an authority figure. Another is degraded to a punitive target of odium and opprobrium. One part of humanity is exalted to the status of righteous gods. Another is execrated to the status of rotten devils. By polarizing the human condition, punishment becomes the tree out of which grows the baleful knowledge of good and evil. Punishment is not the *consequence* of eating of the forbidden tree – it is the primary *cause* of it. What if Yahweh, instead of banning the tree of knowledge of good and evil, went to the root of the problem and prevented the seed of punishment from being planted in the brain of man? There wouldn't even be a tree of good and evil to speak of, and we would be living as an inclusive and inte- grated species, without penal laws and punitive wars!

Punishment has a bipolar frame of reference. God and devil, superman and subman, are the goalposts of the pu- nitive order. Once this framework is up and running, it becomes a warrant for administering ever more punish- ment – a license for an ever-expanding range of punitive judgments. From the branches of the mythic tree of good and evil are carved many of the tools of punishment, the rods, canes and switches that inflicted pain and shame on generations of children since the Fall. Just as current flows between positive and negative terminals of a battery, so painful, sometimes lethal currents flows between the di- vine and demonic poles of the punitive order. This is where the main body of humanity is required to function as a conductive medium.

After all, the extremes of the punitive order are a mi- nority. Most people are not gods or supermen; nor are they devils or submen. Nevertheless, the violent charge that makes up the current of punitive judgment must pass from the one to the other. High and mighty gods like Jehovah or Jupiter can hurl thunderbolts to squash manifestations of lowlife flat. Wannabe gods like lawgivers, generals, judges, are like Ben Franklins who collect and store the lightning bolts of the gods in institutional jars. As they are handicapped by the usual assortment of human limitations, even with the latest technology they need armies of deputies and technicians – sergeants and bailiffs, wardens and commandos, bomber pilots and drone operators, trigger-pullers and button-pushers – to throw their judgmental bolts of vengeance about. This is where it gets tricky. In order for the machinery of punishment to run smoothly, the army of human agents that serve as a conductive medium can only do their job by getting their personal feelings out of the way.

Through a capacity for feeling and fellow-feeling, we treat others as we would like to be treated – with respect, empathy, kindness. Our heart connection to others is the GPS for following the Golden Rule. You would think if we made a priority of accessing our feelings, it would melt the barriers in the way of coexistence. The problem is if we bridged our differences by linking the emotional centers of our being, it would be a bridge too far for the powers that be. It would nip the impulse to hurt others in the bud and rip up the whole premise of punishment. You would think it is good for people to get all touchy-feely and stop acting ashamed of sticking by each other in gooey ways. Except that it would gum up the works of punitive justice but good!

Plainly the *punitive* imperative to inflict pain on others, and the *empathetic* imperative to feel the pain of others, are diametrically opposed and mutually exclusive. If we just cultivated our inborn capacity to *commiserate* with each other, we wouldn't have to plot, organize and intrigue to such machiavellian extremes to make each other's lives so *miserable*. The *law of retaliation* by which we make others suffer is only necessary in a society that makes it a point of hardened pride to live outside the *law of compassion* by which we feel the suffering of others.

In theory, the magistracy of power can inherit the lightning bolt of gods – and evildoers can be found deserving of being struck down. But if humans *en masse* were to retain their native vulnerabilities, their tender sensibilities, they would be hard pressed to pass electric currents of pain to their fellows and the machinery of punitive justice would suffer a mass power failure! If they could remain grounded in their human kindness, they would tell the powers that be to go fly a kite! For the key to the functionality of a punitive frame of reference is not obedience per se. It is overcoming the personal feelings of mindful resistance that rule out obedience.

That is why the enduring legacy of *mass death* is a mass movement of *emotional deadening*, religiously perpetuated from generation to generation. And why the exercise of power is so inimical to life, liberty and the pursuit of happiness. We think of power in hyper-masculine terms as an ideology of hardness, cruelty, violence. But to be powerful in this sense means cutting out the feeling centers that are the heartsprings of verve and vibrancy. Is this not an act of *psychological castration*? To engage in supposedly manly displays of force and aggression we must be robbed of the touchy-feely connections that are essential to human intimacy. Is this not a form of *emotional emasculation*?

If we are to play rough and act tough, at least let us face up to what has to be surgically removed to carry off this show of machismo. It is not the sex organ by which we achieve physical union. It is the organ of love - the facility for empathy and compassion - by which we achieve spiritual unity. The age-old campaign to beat us into a race of rugged, flinty heroes demands that we mutilate our wholeness and eviscerate our core of vitality. Why else do you think the strength of God is made perfect in the weakness of man? To be worthy of the kingdom of heaven, man must mount a withering assault on everything that makes him the animated, ebullient creature he is, fully alive and free to live out his potentialities to the full. And to be worthy of the kingdoms of earth, man must hollow out his humanity through a regimen of self-abasement and self-betrayal and become a gutless wonder.

As much as we glorify the he-man, no pedestal is high enough to hide the fact he is an emotional weakling, a psychological cripple who lacks the courage to show his true feelings and honor who he really is. Unrelenting pressure has to be exerted to embody the strong, silent, stoic type, for unless we go along with this spineless dumb show, the whole punitive order has no way to project power and strength!

39. THE DEMONIZATION OF BABIES –
THE ENABLING MYTH OF CHILD ABUSE

The cosmic forces of destruction became the source of all evil through a deadly trifecta of developments. *Their personification as gods. The sacred imperative to be akin to gods. And the corollary to repudiate one's own kith and kin.* Survivors identify with doomsday gods like an abused child with his aggressor. Just as the abused self is seen from the abuser's perspective as worthy of the abuse heaped on his innocent head, a victimized humanity is seen from the perspective of gods of wrath as worthy of the destruction visited on its innocent head! The gods come to be seen as righteous judges by damning man as an errant child deserving of punishment.

When this psychodynamic plays out between parent and child, it powers the cycle of intergenerational abuse. When it plays out between the divine parent and the children of men, the ramifications are far-reaching for the history of religion, the progress of civilization and the fate of mankind. As one part of humanity identifies with gods of wrath, the other stands in for the body-battered or earth-shattered child. The celestial forces that spell the destruction of mankind are blind, insentient forces, the raw brute energy of matter unleashed with a big nihilistic bang. The child that heralds the procreation of the human race is a conscious, highly intelligent life form, a product of millions of years of painstaking evolution. As dooms-day survivors no longer seem to know which end is up, the brute, insentient forces of the cosmos take on the aspect of wise, powerful gods, guiding the affairs of men. And the child as the keeper of the human flame, the template of a highly intelligent order of being, becomes the purveyor of darkness and chaos!

This is no mere ordinary reversal. It is a radical inversion, a catastrophic upset – a ***bouleversement***. As post-apocalyptic survivors align with a divine order of being, it is no longer possible to see children as innocent bundles of joy, exquisite packages of emotional and mental intelligence to which we just add love. The gestation of children strikes fear in the parent's heart. Their birth is misperceived as a menace, a specter of destruction and chaos. To save the social order, to preserve the divinely sanctioned political order, parents do the unthinkable – they abandon their children, banish them to the wilderness, or place them under a sentence of infanticide.

It makes sense in a bizarre way. We deify destruction and demonize procreation. The incandescent explosions that fill the vault of heaven are the work of gods. The luminous expressions of life that swells the womb of the earth mother are the devil's work. Praise and adoration are offered up to forces of death and destruction that portend the end of the human race – and alarm and loathing attend the forces of birth and gestation that perpetuate it! Skeptic and cynics who pride themselves on their grasp of material reality should never forget it. The foundation of all the brutal attacks on our innocence and vulnerability as a species is not laid down through a calculus of realpolitik, but through *dreams, visions, and hallucinations.* "Infinite times his mother, amid the visions and delirium of dreams, saw her entrails being burst by a bold *monster* in human shape; dyed in her blood, he was killing her, born to be the human *viper* of the age."[1]

It is official now. The source of apocalyptic rupture is relocated from the rogue forces in the sky to the rogue fetus in the womb! Nothing more strikingly proves the lengths to which survivors go to ally with the celestial forces of doom. Man absolves himself of the atrocity of infanticide through a paranoid projection that makes the child guilty of matricide or patricide! The crooked serpent is confused with the immaculate god – and the squeaky-clean child is mixed up with the viper. Refusing to stay within human limits or respect the body's limits, the child bursts forth like a monster to swallow up the parent! It is the primordial crime. *The fault is not in the dragons that lurk among the stars but in the monsters that lurk below the facade of the innocent child! The actual tyrant is not the monster Leviathan but a potential tyrant* in the womb![2]

Here is the origin of *original sin* – the baby as parricide, entering the world with the sinister intent of shedding blood, not suckling milk. *It's what gets the child expelled from his original paradise – the bliss of being cradled in a maternal universe.* It is a foregone conclusion. As soon as celestial forces of chaos and destruction are deified, children are demonized as specters of chaos and destruction. This is the enabling myth for a whole tradition of infanticide and child abandonment. Beyond that, the arch-perversion of deifying the forces of destruction and demonizing the products of procreation makes it necessary to merge

---

[1] Pedro Calderon de la Barca, **Life is a Dream** [p. 18] Dover Thrift Editions

[2] Pedro Calderon de la Barca, **Life is a Dream** [p. 20]

our identity with disembodied gods by making enemies of our flesh and blood. This becomes the defining model of religious worship, penal law and order, and military retaliation – in a word, the backbone of civilization.

Though it starts with rejecting babies outright, infanticide soon proved to be too drastic a solution, not to say a wasteful, impractical one that could lead to the extinction of the family. So the death sentence children labor under is commuted to a formative life sentence of assault and battery of varying degrees of ferocity, and the brutalization of our young became the norm for thousands of years. If parents make a ritual practice of beating their children, in all honesty it can be said the gods put them up to it. The rod has always been a weapon of punishment wielded by parents as proxies for wrathful gods. Countless parents and teachers salved their consciences of the terrible violence they inflicted on children by clinging to the belief it is God's will. Children are routinely violated and battered on the strength of authority delegated from gods to kings and clergy, from kings and clergy to magistrates, and from magistrates to parents and teachers.

The justification for the rod came not from any factual evidence of evil in our young. It came from a runaway paranoia, a persecution complex that conjured up the specter of the *spoiled* child – the child freed from all human limits, bursting through the natural order, running riot and threatening to bring down the kingdoms of heaven and earth. The rod was deployed to crack down on this chimera of a monster. Violence against children was sanctioned to defend the domestic order, safeguard the social order, uphold the political order, and preserve the divine order.

There may not be a scrap of paper proving Adolf

Hitler authorized the destruction of the Jews, but there is an extensive trail of Holy Writ attesting to a God that authorized a veritable holocaust of child abuse down the ages. Indeed, all the terrible things that children have ever done by claiming the devil made them do it are mere foolish pranks compared to the mayhem and murder that parents inflict on children by claiming God made them do it. Humans have long wormed their way into the good graces of phantom gods by scourging their flesh and blood. From there it is a logically seamless transition to a whole tradition of ascetics who seek to get into God's good graces by scourging their own flesh and blood.

Get it? *Flesh and blood* applies to the physical body as well as to progeny The same punitive wrath directed at the child's body to make him acceptable to God, parents and society, is directed at one's own body to purge it of evil inclinations and make the soul acceptable in God's eyes! In effect, the ascetic champions the cause of infanticide as the official voice of the child who rejoices in the horror of being abandoned to the beasts of the field. "Leave me to the beasts that I may by them be made partaker of God...I should be ground by the teeth of wild beasts, that I may be found pure bread of God." [1]

---

[1] Ignatius, 3rd Bishop of Antioch. *Fathers of the Western Church*, Robert Payne [p. 28]

40. A MODEL OF TEACHING
BASED ON A COSMIC BLUNDER

Today we no longer condemn *infants* to death to preserve the social order. But there are countries that still think it is necessary to sentence *adults* to death to preserve the social order.

We no longer believe that *children* will run riot unless threatened by the punitive rod. But we still believe that *citizens* will run amok unless threatened by the penal rod of the law.

So even though infanticide and child abuse are outlawed the world over, the animus to punish our flesh and blood hasn't petered out. It has gone on to bigger and better things. We are all members of the body of humanity, but that doesn't stop us from scapegoating people who are as full-blooded as you or I, and cutting off their blood supply of love until they reek of death and decay. An obsession with quasi-divine law and order makes ascetics of us all in relation to the body of humanity. We don't do *penances* any more for the sins of the flesh, but those who sin against society do time in *penitentiaries*. Acting as the punitive arm of gods, we force those of our kind who seem in league with the devil to perform a range of *austerities* that only medieval hermits and masochists could possibly believe is good for the soul – from forgoing elemental pleasures in isolation cells to being under the daily threat of being beaten, debased, and reviled.

The consensus of sages down the ages is we are all integral parts of a common body. All of us share the same body parts. But you wouldn't know it from a law of retaliation that compels us to take Christ's chilling command

237

literally. We imagine our hand is giving us offense and proceed to cut it off. We take imaginary offense against our own eyes and gouge them out.

As for war, it is the auto-immune disorder of the human race, where perfectly healthy organs are attacked by our own system of defenses. The Vietnam War is a perfect example. In making an enemy of Ho Chi Minh, a long-time admirer of the United States, were we not attacking a man whose patriotic desire to throw off the yoke of colonialism made him the George Washington of his people, a picture of whom he kept on his desk – along with a copy of the American Declaration of Independence?[1] Were we not, in effect, fighting our own best selves?

The war that planetary gods waged on earth was a flaming hell, much like the war men wage on each other. The forgotten tragedy is that babies were the first casualties. War is a continuation by political means of the ancient practice of making burnt offerings of children. Our ancestors betook themselves to the high places to drop babies into the fire. Modern counterparts take to the skies and drop fire on babies. "What is the difference between throwing 500 babies into a fire and throwing fire from aeroplanes on 500 babies?"[2] In olden times, omens were the smoking gun that caused newborns to be abandoned in wastelands. Nowadays, the imaginary threat of a

---

[1] William Blum, **Killing Hope, U.S. Military and CIA Interventions Since World War II** [p. 123]

[2] Captain Philip S. Mumford, former British officer in Iran. **Human Smoke**, Nicholson Baker [p. 66]

mushroom cloud triggers the missiles that reduce entire motherlands to wastelands.

Here is why the pantheon of poobahs that play god can never bring peace – *by identifying with them we are forced to split off from our kind and become our own worst enemy.* Here is why the panoply of authority figures that play god can never deliver us from evil – *by merging our identity with them we are expected to do evil to our bodies and the body of humanity.* For millennia they force-fed us the same lie. Our communal welfare and security lies in disowning our feeling centers and obeying the commands of quasi-divine authorities. They make-believe it is disobedience that got man kicked out of paradise, when it is falling in line with the gods that led to the Fall of man. By merging our identity with a superstructure of power, we split off from our integral selves and fall for the poisonous knowledge that one part of humanity is righteous and good, the other wicked and evil. *It is this that made it impossible for us to stay in paradise.*

Where did man learn to become such a pushover for obedience if not from the *blame-the-victim* school of coping? *The fault is not in the stars, but in our own wicked selves that caused the stars to fall on us!* Cataclysms are sent to punish man for departing from God's ways. Ergo, to find refuge in God's favor, man must quit transgressing and walk in the way in which the gods command him. *Now therefore amend your ways and obey God's voice and the Lord will change His mind about the misfortune which He pronounced against you.* Mass death befell our species through the shame of disobedience, ergo, the virtue of obedience is the ticket to the good life. *Please obey the Lord in what I am saying to you, that it may go well with you and you may live.*[1]

The myth of the Fall comes of being blind to the fact that man fell out of his paradisiacal idyll through a mass death sentence passed by doomsday gods. The truth is twisted by framing man's expulsion from a terrestrial paradise as punishment for sin. This trauma-addled twistification causes man to jump to a command to walk the straight and narrow that has no precedence or basis in human nature. Out of this cockeyed misinterpretation is forged a model of obedience that becomes the cornerstone of religious and political order. Pedagogy is about leading children – and how we were led as children is how so-called leaders lead us as adults. From here on out pedagogy is bound up with obedience. The first law of pedagogy is to follow the dictates of authority. And no one remembers that a pedagogical order, designed to combat ignorance and correct mistakes, itself came from a colossal mistake based on an ignorance of cosmic dimensions!

---

[1] The first quote is from Jeremiah 26:13, the second from Jeremiah 38: 20

41. THE MILGRAM PARADIGM
AS A FIXTURE OF HISTORY

The art of pedagogy was learned at the feet of phantom gods who hurled cosmic thunderbolts to teach man a lesson. And their pupils were off and running to get their teaching credentials as hierarchs and theocrats.

In a Babylonian palace, authority figures wore royal robes and bore imperial scepters and faithfully aped the gods by drafting armies and drawing on state-of-the-art arsenals for the latest thunderbolts to hurl at errant fellow humans.

In a lab room at Yale University, authority figures come in white coats with clipboards and pens. Their foot soldiers are drawn from a cross-section of American society – postal clerks, social workers, housewives, teachers, salesmen, engineers, nurses. But the authority figures still behave like hierarchs and theocrats of old, commanding armies of citizens on Main Street to deliver bolts of electricity to punish their fellow humans for erring or losing their way.

And what is it exactly that shocks us about this? That hosts of ordinary folks should forsake their humanity and hurl potentially lethal bolts of electricity at their fellow humans to teach them a lesson? *Or that the heavenly gods they worship and the earthly authority figures they obey have been doing it for thousands of years!*

Much of civilized history seems to be organized by the twisted brain of a *mad scientist* conducting a psychology experiment. A class of authority figures takes its cue from thunderbolt-wielding gods – and a class of sinners or criminals pay for its mistakes by incurring their wrathful judgments. In between a mass of humanity serves as a medium for conveying the punitive will of gods or wannabe gods on high to a class of errant victims below. This is the status quo, and as static as it seems over time, the will to dominate always generates pushback and blowback, as it goes against the grain of human nature. Progress is a constant race to see who proves bigger and stronger – a never-ending game of jockeying for the coveted spot of seeing *who* will be empowered to teach *whom* a lesson. The overarching effect is to push the envelope of punishment to surreal extremes. It is to amp up the capacity to mete out punitive judgments, moving ever higher up the punishment scale until man's ability to deliver thunderbolts of wrath matches the gods.

When Stanley Milgram staged his famous experiment on obedience to authority, he could hardly envision what an apt paradigm it would prove for tracking the arc of history. History is a generator of greater and greater mechanical vectors for delivering shock and awe to correct the sins of our species. The height of irony is how shocked we act that average citizens work their way up to a measly 450 volts of energy to correct their fellow man, when nothing short of *full spectrum dominance* satisfies a vanguard of role models, for whom the sky is literally the limit in terms of the magnitude and intensity of shock and awe they unleash to police an erring species. How can we pretend to act surprised that two-thirds of a cross-section of Americans are willing to go along with *electrocuting* their fellow man, when their leader in the Oval Office boasts of making the decision to *incinerate* hundreds of thousands of people with the snap of his fingers![1]

---

[1] When Truman was asked whether the decision to drop the bomb was morally difficult to make, he responded, "Hell no, I made it like that," snapping his fingers. Peter J. Kuznick, ***The Decision to Risk the Future...***

It is a truism that developing weapons to deliver bigger and bigger payloads of nihilistic energy has failed to make the world a safer place. What's worse, it has done little to curb the incidence of evil. On the contrary, the greater the scale of destructive firepower at our fingertips, *the greater the evil that is conjured up to rationalize it.* That is why when the First World War erupted, the British Propaganda Bureau sprang into overdrive to take a peace-loving, peace-keeping Kaiser, whose country had nothing to gain from war and everything to lose from it, and scapegoat him as the warmonger who was single-handedly responsible for starting the war.

This then served as a pretext for punishing his defeated country with a sadistic treaty that stands as a world-class showcase for everything counterproductive about the punitive principle. The country's ethnic integrity and geopolitical unity is savagely mauled, its economy sacked by astronomical reparations, and its psyche seared with the shame of a mythical war guilt. And when a leader arises to right this terrible wrong – by repealing its death sentence, restoring his country's dignity, and rebuilding her defenses to regain her place in the sun – he is vilified as a madman bent on world conquest, and damned with the responsibility of single-handedly starting another world war! For those who fought in WWII for the noblest of motives, here is the question of the century. *How good is a war that is misbegotten by such an evil travesty of justice?*

Punishment is not meant to deter but to feed a crazed addiction to punishment. Every desperate attempt to militate against its horrors is fresh cause for inflicting more. Armed with the scariest weapons of all, leaders can't stop amassing more to blow up earth many times over! And to avoid looking psychotic they paint a vast swath of earth with the broad brush of *Evil Empire* and gibber that it poses an existential threat to Truth, Justice and the American way. Even after working at it for ages, man cannot make a *real* case for an evil that warrants the apocalyptic fury of gods! A five-star U.S. general let on: "Our government has kept us in a perpetual state of fear with the cry of grave national emergency. Always there has been some terrible evil to gobble us up if we did not blindly rally behind it by furnishing the exorbitant sums demanded. Yet, in retrospect, these disasters seem never to have happened, seem never to have been quite ***real***."[1]

The same hyperactive, hysterical imagination that *justified* a war of the worlds by branding man the epitome of evil is hard at work in the 20th century to *justify* world wars by summoning up the bogey of barbaric Huns, power-mad Nazis, evil Communists and the enemy *du jour* – Islamofascist hordes threatening to crash our borders and smash the gates of civilization. Right! *Our way of saving ourselves from a few deranged souls with nothing to live for is to put our trust in leaders who amass enough weapons to commit planetary suicide!* An astronomer calculated "the lifetime probably that a resident of the globe will die at the hands of international terrorists is 1 in 80,000, about the same likelihood that one would die over the same interval from the impact on the earth of an especially ill-directed asteroid or comet."[2] Has no one noticed the national security

---

[1] General Douglas MacArthur, 1957. ***The Pathology of Power***, Norman Cousins [p. 179]

[2] John Mueller, ***Overblown, How Politicians and the Terrorism Industry Inflate National Security Threats*** [p. 2]

state to which we look to protect us from international ter-rorists has already stockpiled enough firepower to lay waste to earth with the impact off a mid-sized comet or asteroid?

Is this what the age-old struggle to appease gods comes down to – capturing apocalyptic wrath in doomsday ma-chines activated by the push of a button? Megalomaniac leaders who play self-sabotaging games of oneupmanship make us believe the greater their military prowess, the more secure our standing. But the greater their military prowess, the more threatening the world must be made to appear to legitimize it. As the capacity for destruction is not innate to man but introjected from cosmic forces, it for-ever outstrips the sum of evil necessary to justify it. Good grows from strength to strength in the circle of life, but there is never enough evil to go around. As the capacity for destruction shoots to infinity, the rationale for it shrinks to nothing. Sparks must be fanned into world conflagra-tions. Petty differences must be inflated into irreconcilable grounds for armageddon. And military establishments that compulsively build on the power to blow up the world must constantly blow up evil out of all proportion to its existence in reality.

42. THE GOLDEN RULE BECOMES
AS SOUNDING BRASS

The destruction of the world is not a response to evil. Evil is a theological response to the destruction of the world. A Day of Wrath when destruction reigns is unconditioned by anything man did. *Unconditional wrath* is the cosmological reality at the dawn of history – and it tends to overshadow, if not eclipse *unconditional love* as the biological reality at the dawn of life.

Unconditionally loved children don't fortify their feelings or armor their bodies. Children of unconditional wrath are at war with themselves. Religion is a protracted struggle to appease the gods who wage *lightning war* on man. This sacred policy of appeasement makes quislings of humans who have to *collaborate* with gods by *betraying* their sensitive, vulnerable side and becoming preternaturally hard enough to hurl lightening bolts in their own right. The *kratos* that is the suffix for *government* signifies the *hardness* that is the signature of *realpolitik*. It takes a warrior's *hardened* mindset to identify with the gods who made war on the world. The heart has to flatline to serve as a clear medium for transmitting the punitive violence of sacred/secular authorities who play god. The functional goal of the Milgram Paradigm is to measure *hardness* or *toughness*, not *obedience* – how *tough* or *hard* does one have to be to inflict a world of hurt.

Courage comes from the word for *heart,* but while we all have the heart to love, none of us really have the heart to kill. The ascetic piety that hates the body enough to make war on it comes more from rupturing the heart than corrupting the body. Today we view the self-loathing, self-lacerating antics of ascetics as cultural critic H. L. Mencken did in his **Treatise on the Gods**: "It is their fate to live absurdly, flogged by categorical imperatives of their own shallow imaginings, and to die insanely, grasping for hands that are not there." To be honest, is a *warrior* mindset any less of a metaphysical absurdity than an *ascetic* one? How is demonizing another body that is just like ours less illogical than demonizing our own body? How is a militaristic agenda to kill the children of Adam any less mystical than the monastic agenda to kill the Old Adam?

It's no more rational to shed blood for geopolitical reasons than for religious ones. It's no less crazy to shed the blood of humanity to prove a love of country as it is to shed one's blood to prove a love of God. Food blockades that emaciate millions of women and children betray the same self-hating, self-denying sensibility as ascetic fasts that reduce the body to skin and bones.

The *mort* in mortification is a synonym for death. Children are the first ascetics to *deaden* their feelings and *die* to their selves – if not for the sake of a religious-industrial complex under the wing of the Church, then for the sake of a military-industrial one under the thumb of the State. History is defined by a Milgram Paradigm because our feelings are a source of shame and have to be routinely mortified to merge our identity with a power structure ruled by pseudo-gods.

A priesthood in sacred robes stands over man, liturgically exhorting him to deliver a daily salvo of withering taunts and abusive strikes to shock his flesh into insensibility. Fire and brimstone once hurled at the body of humanity are reformulated as the sticks and stones man hurls at his body. It's like a chain reaction where the guy abused by his boss goes home and takes it out on his other half, his child, his dog. Gods justify their attack on man by making him out to be wicked – and man justifies the attack on his body by making it out to be corrupt! And from the individual body the chain of violence is brought home to the body of humanity. How love your neighbor as yourself when you have to whip the self by whipping up hatred and contempt for it? How honor creatures of flesh and blood like yourself if you must scapegoat your flesh and blood as a repository of sin before steeling yourself to *aggress* against it to get into God's good graces?

It's here the Golden Rule turns to blood and iron. It is fine to treat others as you like to be treated – *if* how you like to be treated is with care and understanding. What if you believe the right way to treat the self is with disgust and contempt? *Then that is how you treat others.* And the ascetic drive to *crucify* the flesh turns into a *crusade* against integral parts of the body of humanity, demonizing and attacking *them* with a vengeance!

No wonder medieval Popes called for a series of wars against Arabs known as Crusades – ostensibly to recapture the Holy City from people of the same Book as Christians. Europeans with a sense of history cringed when a modern President launched a war on Muslims he bluntly called a **crusade**. Are we still ringing that cracked old bell? Are Muslims as worthy of vilification today as Arabs a millennium ago? Or is it that having once weaponized the wrath of gods, the powers that be must restock their villains from age to age so that they can target them for expunction without compunction? The more senseless the destruction wreaked by a comet, the more morally compelling must the case be for it. Nothing like a fear of the end of the world to panic *libertines* into flagellating themselves for the sexual freedoms that bring the *wrath* of gods down on them. And nothing like the threat of their lives ending in an apocalyptic fireball to panic *libertarians* into beating up their neighbors and cracking down on the political *freedoms* that bring the *hatred* of terrorists down on them![1]

As a teen who drank from the Kool-Aid stand of an evangelical radio ministry, I knew a little something about buying security with freedom. Decades before terror alerts or dirty bomb threats spurred people to seal up homes with duct tape and plastic sheeting, I would retire to my room every Saturday morning and remain cooped up there until sundown. I sat on my bed, poring over the Bible on my lap for hours at a time. What chilling self-discipline for a teenager! No sunny beach for me. No playing a guitar or singing in a band. No parties with friends. My parents grew quite worried about me. *I was in full-blown survival mode.* I was indoctrinated with the belief the world was coming to an end. God was coming to pour out his vials of wrath. By observing the Sabbath, I was showing God I was among the righteous who deserved to be saved.

It was odd behavior for a teen in the 1960s, but not atypical of eras that lived in the shadow of a cosmic terror alert – an ongoing threat of the Hosts of the Lord coming armed with hellfire and damnation. Shut up in tiny cells that served as private GITMOs, religious fanatics tortured themselves daily. They abjured dancing, partying,

---

[1] "Americans are asking 'Why do they hate us?'...They hate our freedoms." George W. Bush, address to joint session of Congress and the nation (Sept. 20, 2001)

pleasures of the flesh, hewing to a puritanical line that would have done the Taliban proud.

As an abode of the dead, Hell may be dead itself, but its ghost live on in fears of vengeful hosts of jihadis. The apocalyptic firebombs come not from gods but men acting in their name. The image of skyscrapers swallowed up in clouds of hellfire is indelibly etched in our brains. It is a Dantean picture of hell and in the name of survival our reaction is the same. Our horizon contracts in fear and trembling. We shut up integral parts of humanity in tiny cells, cloistering them in impregnable black sites, torturing and denying them the pleasures of life. Who needs Bible-thumpers to terrorize us into conformity when we have the Patriot Act. As medieval souls suffered an unprecedented expansion of the power of God over their lives, we suffer an unprecedented expansion of government. As a pall descends on our daily enjoyments, our minds narrow, our hearts shrink. The chilling effects of censorship still our voices. The light of relaxed joyousness goes out of our lives. We sit tight, submitting to bureaucratic inquisitions, minute inspections of our bodies and belongings. Warrantless wiretappings turn private conversations into confessionals. Mass spying turns ordinary pursuits into witch hunts. Everywhere we go we watch what we say and do. The power of religion is in abeyance, but Lord how we are suckered by the religion of power! The power structure is now the new *ecclesiastical* order to which we cling to save us from the infernal torments of terrorism!

This isn't trading freedom for security. It is trading the freedom to breathe, to live, to be, for the faith-based security of being saved from phantom demons, be it in this world or the next! Scaring the life out of people is no way to save lives. "I wonder whether politicians who are using fear to get themselves elected would stop if they knew the harm they may be doing to people's health. Real physical harm. Making people sick. Perhaps even killing them."[1] Escalating terror threats under the guise of defending us from foreign invaders elevates our stress levels and weakens the immunity that is our natural defense against disease and depression! Americans are worried sick because their jobs are going away. They are worried sick their health care is going away. But hey, the military-industrial complex can rest easy, between the threat of Islamic State is *not* going away![2]

Here's how you know the war on terror is a sham. The mantra trotted out to justify repressive laws and wars of aggression is: *We're saving American lives*. Would that be the same American lives they throw under the bus of globalization? Would that be the same American lives they strip of health care and leave to die cold and hungry on the streets? Would that be the same Americans whose quality of life they undermine with every passing generation, until they are dumped on a Third World scrapheap? How credible is it that the excuse for their crimes against humanity is, *We're saving American lives*, when the mortality rate for infants and youth, the mortality rate from gun violence and drug abuse, the mortality rate for the sick and elderly, the poor and handicapped in this country, is the highest of any industrialized country on earth!

---

[1] "We're Being Scared to Death", David Ropeik, **LA Times**, September 22, 2004

[2] "The terrorist threat posed by Islamic State isn't going away." Doyle McManus, **LA Times**, July 16, 2017

43. TERRORISM – CLEANSING A CORRUPT WORLD THROUGH RELIGIOUS VIOLENCE

Once unconditional love is acknowledged to be at the very heart of human society, there is not a shadow of a doubt. The *central theoretical problem of sociobiology* is not altruism.[1] It is power. Love heals psychic tears through reintegration and communal rifts through rapprochement. *Power lives in a house divided against itself.* The ideal of power is to climb to the top and pull up the ladder behind it. The reality of power is to ordain division, disparity, discord.

Here we are, the most highly evolved species on earth, and we still think we need something over and above and beyond us to keep us on track! Only a breakdown of cosmic order could have shaken loose our faith in our nature and led to that breakdown of *mental* order institutionalized in *kratos, cracy, crazy*. Power exists by inventing a phantom species known as gods and imagining there is no transition, no developmental link, no bridge of continuity between the human and divine. "Between God and man lies a 'crevasse', a 'polar zone', a 'desert barrier'."[2]

Power corrupts and absolute power corrupts absolutely. This bare-bones axiom can be fleshed out as follows. *Power ruptures the integrity of the human continuum – and it grows absolute by making the rupture absolute.* This paradigm of power originates in religion in the form of the

---

[1] According to Edward O. Wilson it is. *The Brighter Side of Human Nature*, Alfie Kohn [p. 183]

[2] Karl Barth, said to be the greatest Protestant theologian of the 20th c. *The Question of God*, Heinz Zahrnt [p. 24]

Milgram Paradigm. Man is obligated to attack his humanity by merging his identity with gods or aligning with a power structure presided over by wannabe gods. The needle moves by ramping up the ferocity of the attack. The Other is made to fall further behind into arrears until he becomes a quivering, blubbering puddle you can wipe the floor with. The role of authority is to push the experiment to extremes. It is to multiply the mistakes to magnify the punishment. It is to heighten our detachment from our heart-centered nature to deepen our identification with a harsh, overbearing god. Mounting pressure is brought to bear on subjects from a *polite request* to continue, to making it a *requirement* to keep going, to stressing it is *absolutely critical* to carry on, to declaring it is m*andatory* or *compulsive* to press on to the bitter end. To paraphrase a President, once you start the business of judging, punishing, killing, *you just get deeper and deeper.*[1]

Religion is the mother of autoimmune diseases, tricking the human body into attacking itself, making an enemy of the flesh by fighting its healthiest instincts. In the end, we grow absolutely inflamed against our kind, falling victim to a syndrome of inflammatory responses codified in the theory and the practice of Hell. The object of religion is not to recuperate from a maternal breakdown by mothering us back to wholeness. It is to push man to extremes of dissociation. It is to exonerate gods by excoriating men as sinners. Man didn't even fall from paradise for his sins. He lapsed into sin by losing his spot on Planet Eden. So to save himself from the desolation of a wasted planet he follows

---

[1] President Eisenhower, quoted by Mark Kurlansky, **Nonviolence** [p. 142]

the strait and narrow path right into the wilderness! And the sin of debt keeps piling up until it is impossible to repay and he is damned to hell for eternity!

A shift to secularism doesn't reverse the trend. Man incurs a debt of sin by being robbed of an earthly paradise, his *earthright*. The debt of crime he incurs by being robbed of maternal bliss, his *birthright*. Salvation comes not of rebuilding prosocial bonds with fellow humans but cementing sociopathic bonds with pseudo-gods. As a criminal pays his debt to society, society's debt to him mushrooms out of control. As he spirals into the hole, he is buried in a penal hole. As his odious debt balloons beyond hope of repaying it, he is executed with a surge of electricity.

Crime repurposed with a political twist is *terrorism*. Terrorists are birthed by being wrenched from the holiest place on earth – the sanctuary of a loving support system. They are displaced from that verdant patch of mother earth that is their stake in our garden planet. And the received wisdom for dealing with them is to invade their countries and place them under armed occupation! If pennies on the dollar were spent on making budding terrorists secure in their homelands, we wouldn't have to waste trillions on Homeland Security. How do you solve the problem of denying humans their paradisiac birthright by bombing their homelands to wastelands? The answer is self-evident but it's a non-starter. By dint of mounting pressure to take kindly to organizations with godlike aspirations, we act unbearably cruel and paranoid to our kind. By being brainwashed to rally around the punitive, aggressive agenda of authority figures, we withdraw the common courtesies that do justice to the sovereignty of our individual beings.

To stop terrorism we don't update an Orwellian

toolbox. We scrap the Milgram Paradigm. The sin of terrorism comes of expelling innocent people from their idyllic lot in the olive and almond groves of our garden plant. If they dramatize their plight by bedeviling us, or threatening to hurl us into infernos, we don't tell them to go to hell. We realize they are already in hell because it is *insufferable* to live without their *birthright*, their *earthright*. How many ears do we need to hear their cry for paradise? Having been denied it in this world, it's all the more urgent to have it in the next. Yet what religion believes in unconditional love enough to do away with checkpoints at the gates of paradise. Desert fathers fought to get in by indiscriminately scourging the individual body. And jihadis fight to get in by indiscriminately attacking the body politic.

You have a problem with jihadis who kill for heaven in the hereafter? Give them a heaven to live for here on earth! That would restore the better angels of their nature. It would also put paid to schemes to serve the greater glory of quasi-divine structures by prodding people to methodically strike out at each other. It would reverse the thrust to strengthen the hand of authority figures by pushing people to further extremes of self-destruction, suicidal or homicidal.

For it is not terrorism that defeats us but the self-defeating way we deal with it. We follow an archaic blueprint. Remember how survivors dealt with rogue comets that rammed the cosmic edifice and reduced it to a smoking ruin? Unable to process their magnitude of trauma, they repressed the pain of fiery monsters that stalked the skies and displaced it on fetuses floating in the womb. Their children came into the world with the baggage of apocalyptic doom. They threatened bloody murder, stood to run amok, and plunge society into chaos. Comets were *bad news* until survivors made them as gods. Then they dreaded the arrival of babies as *bad news.*

Predictably, *the devil you banish from your garden turns up again in the garden of your son.*[1] How ancients coped with primordial terror threats spelled disaster for our young. Children were thrust into the world as firebrands, *enfants terrible.* Even if there is a grain of truth in it, the solution is clear. We don't recklessly abandon a child that acts with reckless abandon. We take him in arms, enfold him in the bosom of family, fulfill his needs to bond, belong. Love, the center of our world, centers us in our humanity. Our ancestors took the opposite tack. The child is exiled beyond the pale, forced to survive in the wild without the normal restraints or social limits that come from the humanizing presence of others. From such foundlings arose a class of beings who identified with gods to serve as purveyors of the random, senseless violence of the universe.

Survivors can be forgiven for being too dumb for science and too numb for a science of human relationships. The same can't be said of their descendants thousands of years later. We have a mature science, a maturing science of human relationships. Yet like our ancestors we are gripped by the *power of nightmares,* held hostage by the *politics of fear!* We paint vast segments of humanity with a broad brush of terror. We quake in fear of entire civilizations headed for a tectonic collision with our own. Like mothers of old who gestate monsters they fear, we spawn the monster of radical Islam that turns on us with a vengeance. Radical

---

[1] A paraphrase of a saying by the Swiss pedagogue Johann Heinrich Pestalozzi (1746-1827)

imams, mullahs, ayatollahs that stalk the landscape, thundering threats of fire and brimstone, are a deliberate creation of U.S. foreign policy – a mutant ideology fostered, supported, organized, funded by Washington.[1]

Any researcher worth his salt knows the war on terror is a full-court press for tilting at windmills. The ogreish brood of Al Qaeda and ISIS march under banners woven of whole cloth. Even if there is a grain of truth to it, the solution is clear. We don't mobilize engines of death against those with nothing to live for. We don't occupy the lands of peoples whose stake in an earthly existence is tenuous enough. Ground terrorists in love to stop them blowing themselves sky-high. Make them feel like they matter, like their lives have *gravitas* and meaning. Restore their dignity, reunite them with loved ones, attend to their creature comforts, embolden them to pursue their passion. Their hopes of paradise wouldn't blossom in the smoldering rubble of our landmarks if we wanted nothing more than to help them make their deserts bloom into paradise.

Terrorism is the closest thing to the Apocalypse our ancestors faced, and with all our enlightened progress, the best we know to cope is to wage apocalyptic war on terrorists. We burn their cities, pillage their homelands, kick down their doors, kill their loved ones, and torture them into brute insensibility, until, in the end, *they become perfect conduits for gods of wrath and justify their senseless, random destruction by telling themselves they are cleansing a corrupt world*!

---

[1] Robert Dreyfuss, **Devil's Game, How the United States Helped Unleash Fundamentalist Islam** [pp. 1-2]

44. THE TIMELESS POWER TRICK – SOW A DIVISIVE SPIRIT, SWEAR TO UPHOLD THE PEACE

All our troubles come of not leaving ill enough alone. Man couldn't accept the violence of the cosmos as an accident. He had to frame it as the punitive pedagogy of gods. Religion officiates over the *unholy union* of senseless violence and punitive pedagogy. The fruit of this union is a Milgram Paradigm that reframes in pseudo-moral, pseudo-legal, pseudo-scientific terms, the model of gods who hurl bolts of wrath to electrocute a wicked race. It is the sacred template of a civilization whose authority figures use thunderbolts of gods in one form or other to teach man a lesson. And terrorism is the shadow side of a religious faith that exalts the senseless destruction of human life as an act of divine vengeance. Gods used fire and flood to cleanse a corrupt world. Terrorists use bombs and explosives in the name of God to punish a sinful society or strike down a decadent civilization. *No problem, however, can be solved from the same level of delusional thinking that created it.* The myth of terrorism originated in a culture of trauma harking back to a terrifying war of the worlds – and the last thing that is ever going to solve it is a *war on terror.*

In fact, there wouldn't even be such a thing as terrorism, much less a war on terror, if the Milgram paradigm were not hardwired into the civilized brain and canonized as public policy. If you want to belong to the power structure, you are *required* to do as you are told. If you want to keep your job, bank account, mansion, pension, it is *absolutely critical* to follow the lead of authority figures. If you want to live a long life and not keep looking over your shoulder, it is *compulsory* to keep your head down and not blow the whistle on superiors. It is your duty to do as ordered even if it causes a world of hurt. If you must deny people life, liberty and the pursuit of happiness, your don't demonstrate or remonstrate against it. You issue parking tickets. You evict families. You garnish wages. You slash budgets for free clinics, school lunches and Meals on Wheels. Above all, you act as the eyes, ears, and hands of lords of finance who loot the common wealth and leave people to be crushed by tanks or curl up and die penniless and ill in street encampments. You function as an obedient cog in a military-industrial machine that drives the world to war. And if tens of millions have to die for real, so be it. *So he's dead. I did my job!*[1]

How rationalize doing a job that does others in? *Play the mythical pedagogy card!* You punish people to death to teach good from evil. (And you thought it was the other way around!) Want your tax dollars to go for doomsday machines? Tell yourself they're not senseless weapons of mass destruction but pedagogical tools to teach the peoples of the earth a lesson. Want to advance on the Mohs scale of emotional hardness without any foot-dragging? Tell yourself people walk all over a softie. Want to murder? Tell yourself you don't want anyone getting away with murder! Want to do something bad? Tell yourself you're keeping someone else from doing bad!

What about that protest that rises in your throat? Make sure it is still-born by convincing yourself the situation **requires** you to continue. Or that conscience that pricks

---

[1] Pasqual Gino, Water Inspector, Experiment 7. **Obedience to Authority**, Stanley Milgram [p. 88]

you? Blunt it by arguing that disastrous punishments are **critical** to the social enterprise. Act like sublethal or lethal punishments are **compulsory**, for if you let up on a retaliatory system geared to inflicting greater and greater harm, the civilized experiment would fall apart! How reminiscent of a military-industrial complex that compulsively warns of the gravest consequences unless the American people spurge to close the latest missile gap or spring for the next generation of fighter jets!

This is the ulterior thrust of the Milgram Paradigm. To push the man on the street to go along with an arms industry designed to inflict ever-mounting levels of shock and awe on his fellow man – and to prevail with thinly veiled threats and disinformation campaigns that imply that *it is of critical importance, a national emergency, a matter of life and death that he play along*!

In any case, escalation goes with the penal territory. *Capital* is wed to *punishment* from the get-go. In the Bible, sin is a *capital* crime. *Sin is rebellion against God; it is a traitor's act who aims at the overthrow and death of His sovereign.*[1] For the progeny of doomsday survivors, "the greatest crime of man is being born."[2] And the penalty for this capital crime is infanticide.

If a victim of punishment dies, assume he is irredeemably *evil*. If he survives, assume he is punished *for his own good*. Correction aims to make a better person. Unless the only good evildoer is a dead one. *The evil of sin consists in its being the fully willful rejection of God – an attempt as it*

---

[1] John Henry Newman, **Parochial and Plain Sermons**, V, 1843. **World Treasury of Religious Quotations** [p. 934]

[2] Pedro Calderon de la Barca, **Life Is a Dream** [p. 4]

were to annihilate God.[1] A baby incurs infanticidal wrath for the crime of plotting to murder his parents. Man incurs genocidal wrath for the crime of plotting to murder his gods.

How neatly the truth is flipped on its head! A parent is not guilty of rejecting a child – the child is guilty of rejecting a parent. God is intent on annihilating a species that is bent on annihilating God! As soon as a rage for punishment becomes homicidal, the Milgram Paradigm kicks it up to the ultimate notch of self-justification – *self-defense!* When a punitive judgment climaxes in death, it is not *punishment* – it is *self-preservation.* You can take issue with the unfairness of punishing an innocent baby or a species but you cannot argue with the law of self-preservation.

What a bum rap for humanity! Men have to be *enfants terrible* to justify the terrorism of gods of wrath as *counterterrorism.* Once potentates take on the persona of doomsday gods, the only credible way to justify their death-dealing judgments is by casting man as a wild-eyed assassin, or is it an assas-*sinner,* who plots society's downfall. Once the cult of *kratos* is in place, the power structure gets away with channeling the judgment of wrath-crazed gods by making man out to be the homicidal maniac! Gods or wannabe gods wielding weapons of mass death are no longer the psychotic ones. They are guardians, saviors, protectors. The psychos are the ones running around in our midst. They the *natural-born killers* who are a menace to society!

So gods of wrath become rock stars of mankind, their doomsday judgments enshrined as the law of the land. And the innocent child of an innocent species is born

---

[1] Bruno Webb, **Why Does God Permit Evil?**, 1941

under a evil star! Under the circumstances, it is a wonder our prosocial nature emerges. And no wonder our antisocial power structure endured for so long. *We keep falling for the same old trick.* First, we are born with the stigma of being deicidal sinners, natural-born killers who threaten to go on a rampage, if we are not shut up in a cell or coffin. Then, we are denied our birthright of bliss, throwing us off balance, blighting our good nature, blasting our integrity, raping our innocence, until we are *mad* enough to act out all the *bad* things imagined of us! The acid test of power and authority is to see how far we can be pushed to go to war with ourselves and one another. When we cry, *Stop, we want our planet back, we want to reclaim our unity as a species*, they smugly demur. We *can't pull up stakes and leave! Civil war would break out! Man would fight man to the death!*

Isn't our double take long overdue? If the powers that be weren't busy pitting us against each other, there would be no demand for them to protect us from one another. The strategy has always been to sow a divisive spirit and grow a mandate to uphold the peace. You have to think the worst of human nature to put your faith in an aristocratic order that fancies itself the best. Man has to be raised to be a wolf to man to give credence to the myth of the Good Shepherd.

VI. POWERING DOWN CONFLICT AND WAR

The political state receives is sustenance from conflicts – real or contrived – conflicts born out of State-induced fears designed to solidify herd-impulses into a consensus for social rule by political institutions...Discord, then, is the lifeblood of all political organizations. Rather than eliminate conflict, the State must encourage and promote threats against which it can mobilize and control its own populations. Though its public relations image is to the contrary, the history of the State has been not one of conflict resolution, but of conflict management.

— Butler D. Shaffer, ***Calculated Chaos*** [p. 114-115]

45. GOOD VIOLENCE CANNOT EXIST WITHOUT BAD VIOLENCE

The Rosetta Stone is a slab of basalt discovered at the turn of the 19th century. Inscribed in Egyptian and Greek, it uses three different scripts – hieroglyphic, demotic, and Greek. Armed with these parallel translations, scholars could finally solve the riddle of hieroglyphics and get down to the business of deciphering a written language that had been dead for nearly 2000 years.

The asteroids and comets that laid waste to earth are like the Rosetta Stone. They open up the book of history and enable us to solve its riddles by using three interwoven streams of development. A catastrophic breakdown of cosmic order. The emergence of religion as the worship of doomsday gods. And the onset of political order in the form of a cult of **kratos,** or rule, identified with doomsday gods.

Cosmic violence may be senseless itself, but its traumatic impact on the human species enables us to make sense of the language of death spoken by religion and civilization for millennia. We know why violence is the heart and secret soul of the sacred.[1] Now we know why violence is the heart and not-so-secret soul of secular power. *All forms of government are rooted in violence.*[2] Whatever else may separate religion and politics, both breastfeed on the terror of violence. The sacred bonds that bind man to gods – and the social contracts that bind men to rulers or wannabe gods – are renewed and strengthened from age to age by violence.

---

[1] Rene Girard, **Violence And the Sacred** [p. 31]

[2] Emma Goldman, **Anarchism and Other Essays** [p. 7]

That is why world peace has never been the strong suit of religion or politics.

Religion was birthed in a cataclysmic breakdown of cosmic order, an apocalyptic war of the worlds, and it has never lived it down. Stress, dread, danger are good for splitting people off from their integral beings and aligning them with gods. That is why there are no atheists in foxholes. And why a biblical rapprochement with God requires a relapse into Doomsday – a convulsive descent into world war, or Armageddon, complete with falling stars and melting mountains.

As the political order of *kratos* derives from a belief in gods, it is governed by the same skewed logic as religion. The chief difference is the portentous threat under which the political order flourishes is the imminent breakdown of *social* not *cosmic* order. The collapse of the cosmos – gods going crazy, burning the world – is the dreadful stressor that is the mainspring of religious authority and of such barbaric institutions as initiation rites and human sacrifice. The collapse of society – men running amok, setting fire to the world – is the dreadful stressor that is the mainspring of political authority and of such barbaric institutions as capital punishment and war.

Once a cosmic breakdown gave rise to a belief in the gods, an ongoing fear of *universal chaos* is necessary to sustain a faith in them – and a catastrophic threat of *social anarchy* is necessary to maintain a belief in the political authority that derives from a faith in gods. ***This is the dismal legacy of a Day of Doom.*** A specter of doom has always been the holy ghost of politics and religion. A fear of universal chaos is as crucial to a belief in gods as it is to the guardians of wealth and power who play god. The still current fear of people going crazy with rage may be as baseless a superstition as the archaic fear of gods going mad with wrath. That never stopped hierarchs of old from claiming to be the only thing between us and the specter of gods going crazy and setting fire to the world. No more does it stop rulers of today from claiming to be the only thing between us and the specter of people going crazy and burning down the edifice of society.

In our time we witnessed through our television screens the horror show of wrath-crazed people deploying airplanes as bombs to demolish the Twin Towers. Now extrapolate backwards to a time when celestial forces, personified as planetary gods, went crazy and hurled fire to bring down the edifice of nature. For survivors it wasn't the closest thing to an Apocalypse. It **was** the Apocalypse. This apocalyptic catastrophe did not just shatter the bonds that bound families and communities together. It laid down enduring deposits of ur-trauma in survivors that shattered the integral bonds that hold body and soul together. Our ancestors had no therapeutic means to heal from their shattered planet, their shattered communities, their shattered nerves. To cope with this ur-trauma an artificial form of binding known as *religion* – from *re*, again, + *ligare,* to bind – bound man in allegiance to the very gods who were the symbolic agents of world destruction.

It was a fateful development but a thoroughly unnatural, not to say, preternatural one. It voided the laws of human nature, and elicited pushback as a matter of course. Humans exist to be grounded in their humanity and in earthly nature. Every healthy impulse in them pushes them to restore natural bonds – with each other, with human nature, with the rhythms of cosmic nature. *The perennial challenge for religion, if it was to prevail over time, was to preserve the bonds that bound man to gods.* And it could only do that by conjuring up the specter of universal collapse that shattered the bonds of humanity in the first place, and forged its unnatural bonds with gods.

Whatever hopes we have of a better world must take this reality as the jumping-off point. *The end-all and be-all of religion is to live up to its name by preserving the obligations that bind man to gods.* To this end it invokes the wrath of God as its starting point, instills the fear of God in man, and holds the specter of mass death and destruction over man's head till the end of time.

Not peace of mind but *dread*, not security but *danger* and *desolation,* draws man to God. Not plenty but *privation*, not success but *debacle* and *disaster* sends man running to God. Sometimes you have to wonder whether a violence-wracked society isn't the best tonic for religion. If there are no atheists in foxholes, why not just root out all doubt and disbelief in God by having every refuge of humanity come under fire? Why not convert the last atheist into a god-fearing believer by turning the whole world into a bloody war zone where *everyone* cowers in foxholes?

The heyday of religion is past, but its legacy will haunt us as long as authority figures are in the habit of playing god. The universal dream is for people to follow their bliss and live in peace. The showstopper is not the bad guys. It is those who insist on governing the affairs of men like wannabe gods, armed with conceits of supremacy and world domination. These are the people who swear by a religion of power. And they proselytize for their religion by making-believe *the social order is as fragile and unstable as the cosmic order was for religious man, perpetually at risk of falling apart into chaos.* The religion of power commands widespread belief by promoting the belief in a class of rogue elements who personify gods of wrath, lying in wait to loose murder and mayhem, death and destruction on the social order, unless brought to heel.

One part of humanity is exalted as a priesthood of royals who mete out righteous judgments as wannabe gods of wrath – and they get away with it by demonizing another part as children of wrath who invite those righteous judgments by threatening to go crazy and destroy society. In short, as long as the religion of power fathers *gods* of wrath, it cannot do without *children* of wrath.

Violence reigns because we cannot bring ourselves to issue a blanket condemnation of it. We argue there is *good* violence and *bad* violence. We debate the merits of *righteous* violence and *criminal* violence. We quibble over *just* and *unjust* wars. *We fail to see that good and bad violence are joined at the hip, and good violence just has to have bad violence to absolve it.*

Here is where our political theory is ass-backwards. We see leaders as proxies of gods, imposing law and order on a chaotic world. The vertiginous downside to this delusion is to see society as inherently chaotic and people on the brink of falling out in an all-out war of all against all! Just as nature has to be seen as teetering on the brink of chaos for people to be immolated as burnt offerings to stabilize it, so society has to be seen as teetering on the brink of anarchy for people to be beheaded, hung, shot, electrocuted, or otherwise executed in order to stabilize *it*.

If humans are anything, they are inherently prosocial. If they are seen as the loving creatures they are, the wrath of gods would lose all credibility. For rulers to carry themselves off as gods, the communal order and harmony that is second nature to man must lose all *its* credibility. Thus the genius for cooperation we demonstrate cannot be trusted as the genuine face of humanity. It must be seen as a facade for undercurrents of madness and rage. Human nature as a veneer behind which blood-crazed beasts lurk is not a statement of reality. It is the vestige of a cosmos whose order and harmony were perceived as a veneer behind which wrath-crazed gods lurk! The power of religion is fueled by fears the cosmic order could fall apart at any moment, and only a priestly elite can keep gods from going crazy. The religion of power is fueled by fears the societal order could collapse at any instant, and only a ruling elite can keep people from running riot.

How can a power structure be the *solution* to conflict when its very existence depends on the *problem* of conflict? How can peace and harmony be engendered by a power structure that is indissolubly married to the shaky premise that the family of man is a volatile compound, ready to implode in an orgy of self-destruction? A Hobbesian war of all against all is a relic of doomsday trauma from a cosmic Armageddon. And if a class of schizoid leaders with delusions of godhead has any hope of prevailing down the ages, it is through a set of crazy-making conditions that leave human beings chronically conflicted and in a state of perpetual conflict with each other!

46. LET THE BAD TIMES ROLL

The estranged, aloof worldview that comes of child abandonment is the making of culture heroes. But how is society custom-fit for culture heroes? Believe it or not, by applying the same abandonment principle to society as a whole. The natural impulse of mother and child is to initiate a love fest at birth. If humans kept this up on an interpersonal level – reaching out to each other with the cry, *Don't be a stranger! Come by here and let us get to know each other! Come by here and let us grow in fellowship and amity!* – we would have a Kumbaya society in a heartbeat. But from the time they are little, children absorb the idea of *stranger danger* by osmosis. In hundreds of petty ways they learn that every man is a emotional castaway, marooned in heart, soul, body, condemned to be an island unto himself. The path of survival lies not in cooperation, mutual support, and neighborliness but in lifting oneself up by its bootstraps and looking out for #1. The role model is Robinson Crusoe not Damon and Pythias. Our home is our castle and we are duty-bound to defend it from outsiders. Suspicion, wariness, mistrust, and alienation are the watchwords of the day. Nobody is safe when man is always playing a game of *gotcha* with man.

This is the broken world where the services of culture heroes are in high demand. This is the sort of lonely, fearful, disconnected society where men must hide their emotional intelligence from each other, as women hide their intelligence from men, and make-believe they are too helpless and dumb to get by without a class of higher authorities to protect and guide them, in the form of monarchs, magistrates, money masters. Raised to keep one another at arm's length, we grow to think we need the long arm of the law to get along. Pressed to remember that love is all we need to work out our mistakes and misunderstandings, we draw a blank in the face of crises and jump to the conclusion we must call the police like a *deus ex machina* to resolve our dramas.

Power is established at the ground level by overriding the human instinct to join in love and friendship. Ours is a society where people languish and fail so that power can grow too big to fail. We pray to a predatory overclass to be saved from an underclass that survives by preying on us. A society where chronic maldistribution ensures there are never enough good things to go around, survives through a police state that ensures that the supply of bad guys never runs out.

There is a natural momentum for power, like love, to go from the local to the universal. Love is by nature *between* and *beside*. It enlarges its circle of likeness or kinship by going from the individual family to the family of man. Power is by definition *over* and *against*. It expands its sphere of control to infinity by contracting its circle of affinity until there is just the apotheosis of the One and no others like it. As power is projected to extremes of transcendence, from tribalism to monarchism, from a country to a world federation of countries, the *other* is objectified to a whole new inhuman or subhuman, bestial or demonic level. Kings grow into emperors and preside over world empires by dehumanizing vast swaths of humankind as *empires of evil*.

The blueprint for world supremacy is set forth in the Bible. Here are the Hebrews, content to worship local gods, and along comes one who aspires to be god of gods. How does Yahweh wrest an absolute monopoly of divine power?

Is it through some theological breakthrough into the nature of godhead? Oh no. It is by threatening Israel with a cataclysmic breakdown. It is by ratcheting up the terror level! *Assyria is transformed into the Evil Empire and brandished like a sword of doom to spook Israel into embracing the One beside whom there are no others.*

The astonishing thing about Yahweh's strategy as laid out in *Isaiah* is its mind-boggling audacity. The divine king's most basic function is to protect his people from foreign threats, yet here is Yahweh delivering his people into the hands of their enemy. To discover that the Assyrians or Babylonians are doing Yahweh's work is like discovering that terrorists who blow up our public spaces are working hand in glove with the security forces that are meant to protect us! This is not a mere dereliction of duty on the part of divine protectors/guardians. It is a flagrant act of treason, a betrayal of cosmic dimensions! It goes to show there is a irreconcilable difference between the PR image and the real function of power. The PR image of power is to protect the people and safeguard their welfare. *The actual function of power is to aggrandize itself and clinch its claim to world supremacy!* And it is not above resorting to treason to accomplish its goal, because there is nothing like threats of annihilation to convince a people to renounce its right to think and feel for itself and fall under the dominion of its avowed protector-guardian!

Here is the biblical precedent for what in modern parlance is known as the ***false flag operation***. Yahweh makes no bones about inviting a foreign power to invade and overthrow his own people. He brazenly acknowledges that the Evil Empire of Assyria is the rod he uses to scourge his own people to make them more pliable to his claims of world dominion. Secular leaders are more circumspect. Their hand may firmly grasp the weapons with which they strike the fear of God into their people, but they take great pains to hide their complicity. The false flag operation as such has people coming and going. The right arm of government makes a federal case out of protecting the people, while the left arm covertly terrorizes them behind the scenes!

When the scales first fall from people's eyes, the sheer duplicity of a false flag operation causes their jaws to drop to the floor. But the logic behind it is bulletproof. If people are left to their own devices, they wouldn't need a power structure to live in peace. A *kumbaya society* may be the butt of jokes but the fact is, if it were a reality, the power structure itself would be a joke. Government is organized with the intent of protecting people, but the stark reality is the more vulnerable a population is to outside attacks, the greater its willingness to surrender its freedom, and the commensurately greater is the power of the state! *There's nothing like the wolf of terrorism at the door for throwing the keys to the kingdom out the window.*

As much as we hope for peace and plenty, it is a forlorn hope under the present order. Each time ancient Hebrews experienced a period of peace and prosperity, they abandoned their belief in a supreme being! "Only historical catastrophes forced them to turn to Yahweh."[1] Why should wannabe gods ever spread the wealth around when by doing so they undermine their pretensions of world domination and embolden people

---

[1] Mircea Eliade, *The Sacred and the Profane* [p. 126]. **Cosmos and History** [p. 103]

to abandon their faith in political structures that monopolize power at the top?

People have a natural tendency to follow their loving instincts to create a society founded on trust and cooperation. The only way to wash it out is with a flood of adrenalin – the raw fear that comes of creating a world of menace, peril, threat. That is why it is such a bad idea to hand a sword to the powers that be in the hopes that it will be used to protect us, because in their hands it invariably turns into a sword of Damocles. Think **Operation Gladio** – the sword of terror that was dangled over Europe by the very organization set up to guarantee its freedom and security. Its wave of terror bombings of train stations and public squares strikes normal people as demented, but to the power behind the throne it is just the thing to drive liberal notions of shared wealth out of people's heads and brainwash them into seeking refuge in the fascist arms of a plutocracy.

If no one recalls Gladio, it is because it is eclipsed by 9-11, the mother of terror attacks and of a whole new generation of bombings from Berlin to Bali and Madrid to Mumbai. People fight and die to be free. *But to be free, people must feel safe and secure.* What a hideous travesty to twist that into the claim that freedom must be traded for security, when feeling safe and secure is a precondition of freedom. The powers that be know that to rein in freedom with repressive *security measures*, they must heighten insecurity to the point of panic and dread. They must put the fear of God into us with horror shows of thunderous explosions, earth-shaking firestorms, tempests of gunfire.[1] Only when we are terrorized out of our minds will we line up like robots and give them the mindless obedience and un-thinking submission they crave. Play this out to its logical conclusion and we go from the germ of a police state to a full-blown one. Just ask the rough beast who was no slouch at making the police state seem like the Second Coming. "Terrorism is the best political weapon for nothing drives people harder than a fear of sudden death."

And you must drive people preternaturally hard if ex-isting laws on the book against murder are not enough and you dream of adding thousands of new provisions to the penal code every year. You must drive people witless with fear if you want to catch every citizen on camera hundreds of times a day, and envision CCTVs, biometric scans, gi-gantic databases, tracking technology and microchipping in every corner of the earth, *until your ability to see and know everything on earth approaches the omniscience of God*. You have to drive people clean out of their minds with terror if you want to rule by executive fiat, assassinate who-ever you please and shoot to kill suspects whenever you please, field drones and laser weapons and wage weather warfare with impunity, *until your ability to achieve full spectrum dominance assumes the omnipotence of God*.

Does an omnipotent, omniscient state make society saf-er? Or doesn't it go on incubating planet-threatening mon-sters to justify its existence? Bogeys have to be conjured up from every crop of babies, since hyping the dangers they

---

[1] "Thou shalt be visited of the Lord of hosts with thunder and with earthquake, and great noise, with storm and tempest, and the flame of devouring fire." Isaiah 29:6

pose is the only way to combat them with wonder weapons from the arsenal of gods. Start with the flawed premise that cops are necessary to keep society safe and you wind up arguing that humans are the ones innately flawed and too broken to take care of each other. As you progress up the Milgram scale, putting increasing stock in police powers and policing strategies, you reach a stage where garden-variety cops are militarized into warrior-cops, and warrior-cops are metamorphosized into the vengeful hosts of the lords of the plutocracy. And the only way to make sense of that scenario is for evil to mushroom to astronomic proportions and bring the hallucinatory woes of the **Book of Revelation** down to earth.

There will be massacres on school buses and in day-care centers, ultra violence by Mexican drug cartels spilling over the border, the systematic ambush, murder, and execution of cops, ISIS-inspired butchery, September 11-scale terror attacks.

And the Book of Revelation wouldn't be quite updated for the new millennium without the latest spin on child-bashing. Biblical children of wrath spiral to whole new depths in a *vicious generation of kids raised on the sickest movies and sickest video games, who will soon give us crimes as adults we never dreamed of.*[1]

A cosmic threat of catastrophe sealed the infant's fate as an omen of disaster – a comet in human form. This surrealistic projection weathered the passage of millennia. Never having came to terms with a past of mass murder, we have to listen to terrorism experts intone *the future is*

---

[1] David Gross, "Professor Carnage", Steve Featherstone, **New Republic** magazine, May 2017 [p. 22]

mass murder.[1] And since we never arrested, much less reversed, the process of conjuring up deicidal sinners and homicidal maniacs to justify a power structure modeled after wrath-crazed gods, we are doomed to live inside the nightmares of those who populate the world with them.

---

[1] Ibid.

47. SOME KIND OF *CRACY*

A catastrophic breakdown of cosmic order may seem like the end of the world, but there is no better starting point for decoding the past. Survivors had no one to help them resolve their trauma and get back to normal. They adapted to a world gone crazy by going crazy themselves. They split off from their humanity and came to believe they were gods incarnate. And they perpetuated this as the norm by making other people crazy enough to split off from their selves. *This is where control comes in.* Control is an inherently crazy-making mechanism that denies us the right to be in our right mind. We are not free to be ourselves but forced to be something other than ourselves. We are cheated out of self-knowledge and made over in the image of those who command and manipulate, cow and coerce us into supporting their delusions of godhead.

History is the story of how this class of wannabe gods, suffering a schizophrenic breakdown of humanity, developed a mania to control their fellow humans by violating *their* right to be themselves and forcing them to kowtow and cater to their grandiose self-concepts. Can it be that the comets that hit the earth left us so spaced out for the next few millennia that it hasn't hit us yet? *Our **social** order bears the imprint of a breakdown of **cosmic** order.* How could we take people torn loose from the moorings of their humanity and turn them loose on posterity? How could such a crazy state of affairs prevail that every time we speak of government – whether as theocracy, autocracy or plutocracy – we came within a hair's breadth of using the word **crazy**!

As a society we are old enough to know that kings as gods incarnate, with direct lines of communication to gods, and special rights as descendants of gods, are insane. As a civilization we are like a young schizophrenic girl only just discovering the two pillars of her pathology – a *rage to destroy* and a *rage to control*. "I literally hated people, without knowing why. In dreams and frequently in waking fantasies, I constructed an electric machine to blow up the earth and everyone with it. What was even worse, with the machine I would rob all men of their brains, thus creating robots obedient to my will alone. This was my greatest, most terrible revenge."[1]

These may be the stark ravings of a schizophrenic girl, but look around. The lack of love for humanity on the part of our leaders is equalled only by their pathological hatred for it. And they prove it every day by making her two sickest fantasies come true! They build killing machines for blowing up the world, going so far as to label their policies of deterrence *MAD*. And even if you can take the *military* out of military-industrial, you are left with an industrial complex that doesn't have to destroy bodies because it does a bang-up job of destroying minds. Instead of inventions like mechanization and automation being used to free humans from drudgery, they are used to rob them of their dignity and personhood, souls and minds, by treating them as cogs in a machine. And instead of an invention like money being used as a medium of exchange to make life easier, it is used as chips in a faustian bargain, requiring people to sell their freedom, creativity, peace of mind, sanity, and their very souls to survive and thrive in society. Thus *Modern Times* become a cruel caricature of progress, epitomized by a Depression-era character driven crazy by

---

[1] Renee, ***Autobiography of a Schizophrenic Girl***, Foreword by Frank Conroy [p. 47] Signet Book

a system that uses and abuses people without regard to their welfare and happiness.

How do they get away with it? How is the legacy of the emperor with no clothes successfully preserved in a system of imperialism that stages brutal interventions in century after century, denuding country after country of its wealth of natural resources, blocking the fulfillment of its indigenous needs, and stunting its freedom to chart its own growth and development? It is done with a propaganda ruse that inverts reality. *Madness is covered up with a veneer of sanity – and sanity with a veil of madness.* Here is how it works. Humans are designed to live by the laws of their nature. When we are not free to live according to our lights, we go over to the dark side. Conversely, we must emerge from nonage and come into our own to partake of enlightenment. A propaganda trick flips the truth on its head. *The imperium of control and domination is confused with an enlightened order of civilization. And when you throw it off and emerge from nonage, it is not to become enlightened. On the contrary, it is sink into a dark ages!*

This is the confounded trick that keeps the whole crazy show going. Tyranny, regimentation, coercion, world domination – these are not *obstacles* to emerging into a state of enlightenment, where people are free to use their minds and become their own persons for the first time. They are bastions of civilization, bulwarks of a *Pax Romana, Britannica, Americana* – and their collapse is lamented because it heralds the onset of a dark ages, a period of chaos and anarchy!

What makes this more than a mere semantic switch is a self-fulfilling dynamic that reinforces it with a degree of verisimilitude. The control and regimentation that is a function of *crazy* is inherently *crazy-making* – and needless to say, you can't be driving people crazy without forging a perfect rationale for imposing even greater forms of control and regimentation. In other words, if people aren't crazy to begin with, you can certainly drive them crazy by *controlling* them in all kinds of abusive ways that cause them to lose the ability to exercise their own minds. And then you are justified in putting them in a straitjacket and committing them to an institution.

Contrast Milgram's experiment on obedience with R.D. Laing's study of the families of schizophrenic children. Children are denied a mind of their own. Mothers never let them have their own ideas. Fathers never let them live their own lives. Children are forbidden to express their feelings and be themselves. When these children, against overwhelming odds, break out of this oppressive state of nonage – when they strive to express their feelings, follow their initiative, explore their own ideas and assert the right to have a mind of their own – they are accused of lapsing into *mental illness*! It is not crazy enough that they are controlled and regimented in all the abusive ways that cause them to lose their minds. When they strive to break free of a home that straightjackets their emotions, and regain their minds for the first time, they are damned as *crazy* indeed. And of course, what do you do with crazy people? You straitjacket their physical movements and lock them away in institutions that are even more oppressive than their homes!

It would be one thing if these sorry excuses for parents were an anomaly. But they are role models for institutional embodiments of power. From olden times the institution of kingship was held up as the norm for society. It was the sacred talisman against the scandalous possibility that humans might emerge from nonage to discover they are free to think and feel for themselves. *God forbid we should return to those days when there was no king in Israel, and everyone did what was right in his own eyes.* That way lies anarchy and chaos! So when *obedience* is enshrined as the norm, it is of necessity *blind* and *mindless*. Nowadays people don't have to be kept in chains of ignorance and illiteracy – not when they can be *dumbed down* by an education system that teaches a Disneyfied view of history and government, and *numbed out* by a political system that appeases the hunger for change with cyber circuses and Wonder Bread for the soul.

Humanity is prey to many superstitions, but surely the *most* grotesque one is this. The doomsday gods who terrorized us out of our ever-loving minds deserve to be revered and worshipped because they are the only ones qualified to bring sanity and order to humanity! As the absurdity of this logic is inherited by the political dimension, it begets the *second* most grotesque superstition. The mortal kings who are insane enough to believe they are gods deserve our allegiance and homage because they are the only ones qualified to bring order and sanity to society!

Even as astute a student of humanity as England's premier playwright conceived of kingship as a massive wheel that cannot come off without pitching society into wholesale ruin. Obviously he didn't live to see the British monarchy reduced to a tourist attraction. Nonetheless, the magic power he attributed to the "wheel of majesty" we continue to attribute to the "thin blue line" that reputedly divides civilization from chaos. If the punitive arm of the state were cut off, we are warned in alarmist tones, people would run amok and society would go to rack and ruin!

Like all such ghost stories, the origin for this one is religion. The prototype of the spiritual and secular head is God, and it is of God it is said that if he didn't exist, everything would be permissible. "I remember a fellow student in my college days, an ardent Christian, who told me that if he did not believe in a future life, in heaven and hell, he would rape, murder, steal and be a drunkard."[1] The speaker calls such a person a *sham civilized being*. "Not only could a Huxley, a John Stuart Mill, a David Hume, live great and fine lives without any religion, but a great many others of us, quite obscure persons, can at least live decent lives without it." In fact, the largest study to date of religion and family life discovered high levels of family solidarity and emotional closeness in secular households. Apparently, godless youth can display strong ethical standards and moral values without the help of mealtime prayers and morality lessons at Sunday school.[2]

To think we need cops on the beat to keep humans good is as ludicrous as thinking we need a glorified policeman in the sky to keep them from doing evil. People free to own their feelings and be their true selves do right by their fellow man as a matter of course. The idea that human nature will run amok without punitive checks is like the old wive's tale that without swaddling bands infants will bite off their fingers and scratch out their eyes. It is the act of

---

[1] Walter T. Stace, *The Atlantic Monthly*, Sept. 1948 Melvin Rader, **The Enduring Questions** [p.507]
[2] "'Godless' Kids turn out just Fine" – Phil Zuckerman, **Los Angeles Times**, January 15, 2015

intruding, interfering, in others' lives that denies them the right to be themselves and drives them beside themselves with anger and frustration. It is forcing people to do things they hate, to live lives of quiet desperation that make them dread getting up in the morning, that causes them to to go mad and suffer a psychosocial breakdown. It is this that gives a veneer of truth to the myth that humans will go crazy without the fear of external authority to keep them in line. It is this that produces outbursts like the Murray-Hill riot when the Montreal police force went on strike.

As it is in social relations, so it is in international relations. "There is, as far as I know, only one certain rule in international relations. Interference by one country in the internal affairs of another causes resentment. It is sure to produce a result exactly the opposite of that intended." These are the words of Allen Dulles who, along with brother John, was responsible for crushing the freedom and autonomy of more Third World peoples than any pair of imperialist siblings since Romulus and Remus. When Iran just wanted to be left alone to harness its resources for the good of its people, the brothers saw red – and Mossadegh, the democratically elected leader who championed his people's need to reclaim their own energy resources, was branded a madman by John who, along with brother Allen, went about arranging to have him put away.

Here the Dulles brothers enter the clinically deranged territory mapped out by Laing. To guardians of this sort a child is ill for daring to be her own person. The problem isn't the child but the parents who wind up driving the child crazy. A fancy name for crazy-making guardians is *schizophrenogenic*. Labeling a popular leader like Mossadegh a madman and intervening to place him under house arrest, the Dulles brothers made an entire people mad at America for its oppressive legacy of subversion. It's nothing new for Americans to plaintively ask why terrorists are mad at us. In a National Security Council meeting, just months before the ill-fated Iran coup in 1953, the man Americans had elected President on the strength of his campaign slogan, *I like Ike,* cluelessly wondered why we can't "get some of the people in these downtrodden countries to *like* us instead of *hating* us." *How about the black record of U.S. interventions around the world, spearheaded by the likes of your crazy-making administration, Mr. President Eisenhower!*

As a pretext for the Vietnam war, the domino theory was a bust. But boy, did the U.S. coup in Iran, codenamed Operation Ajax, cause the dominos to fall with a vengeance! Poisoning U.S.-Iran relations for decades, it led to the embassy hostage crisis and U.S. support for Iraq in its internecine war with Iran. It strengthened the most radical, reactionary elements in Iran and whipped up a fierce anti-Western campaign that financed and armed Hamas and Hezbollah, sent agents abroad to kill Iranian dissidents, and bore infernal fruit in the 1983 suicide bombings in Beirut and 1996 attack on the Marine barracks in Saudi Arabia. Its model of religious fascism even served as the inspiration for the Afghans who founded the Taliban. In the end, a persuasive case can be made for extending the line of toppling dominos from Operation Ajax and the Shah's repressive regime through the Islamic Revolution and the made-for-TV collapse of the Twin Towers on 9-11.[1]

---

[1] Stephen Kinzer, **All the Shah's Men, An American Coup and the Roots of Middle East Terror** [pp. 203-204]

48. TO CUT AND RUN FROM ARMAGEDDON

Perseverance is an admirable trait, but it does beg the question. *Where does it get us?* To *hang tough* and *stay the course* sounds noble – unless we are on a course for Armageddon. From the outset it is clear. Splitting off from our humanity to personify gods of wrath is not a coping strategy that bodes well. And the ghastly evidence piled up across the centuries bears out the worst. Staying the course mapped out by the Milgram Paradigm leads to waging apocalyptic shock and awe on our kind. To reevaluate our course leads to a foregone conclusion. Cutting and running is the only sensible thing to do, if we want to stop escalating a war of the worlds into Armageddon. But each time we reach that decision, two things jump out to grab us by the throat.

First, there's the threat our fellow man will go crazy and wreak havoc. If we dare fold up our tents and shutter our Abu Ghraibs and Guantanamos, the heavens will fall and murder and mayhem will be loosed on the world. *Always with the doomsday scenarios!* To force people to rethink a natural aversion to violence, just dangle the picture of their wife being raped or children murdered. *Breathe not a word about what would drive someone crazy with fury, grief or despair to ever want to do such a terrible thing!* You wouldn't want the background to turn out to be just like the foreground. You wouldn't want the rationale for retaliation to fizzle out because every crime is already an act of retaliation for a prior one. Oh no, best not to look into what makes someone mad enough to attack your family, for all you will find is that nothing unhinges people more than the loss of childhood security that comes of watching *their* families ripped apart by neglect and abuse, poverty and war!

And then there's the *shame* of not being tough enough to power through. In a civilization married to the Milgram Paradigm for ages, cutting and running carries a whiff of *cowardice*. It is fiendishly clever, you have to admit, to brand the capacity for fellow-feeling a *character stain*...to decry the hunger for peace as a *badge of shame*...to identify the need for intimacy or solidarity as a stigma for which we should be shunned, taunted, and rejected beyond the pale, if not of *mankind* then of *manliness*. That's why we stay the course and follow the Milgram Paradigm to its bitter end, even if *kills* us. Or makes us an authority on killing! No one in his right mind would go the distance to become a *killologist*, but paradoxically we do, because we are terrified that the shame of cutting and running might kill us. With all the lip service paid to love, no one speaks out as an authority on love, because the shame of being a *loveologist* might destroy us!

These are the twin deadly threats that make us persevere to the edge of doom. The threat of *physical* death keeps us on the straight and narrow path to Armageddon. The threat of *psychological* death keeps us from returning to the ground of love. It is a one-two punch. The first threat takes the physical form of the murderous rape of our family. And the second takes the psychological form of the shame of losing our manhood and being made to look effeminate or effete.

Think of the first question Governor Michael Dukakis was asked in the 1988 presidential debate. "If Kitty Dukakis were raped or murdered, would you favor an irrevocable death penalty..." Ever wonder how low the state would stoop to justify its death grip over us? Here is the answer. To justify its life-and-death powers over us, the state would have us imagine our family ravaged, and our loved ones slaughtered before our eyes! God forbid we should be asked to envision everybody in our species sharing the same genetic make-up, or being part of one big happy loving universal family – *because then the power functions of the state would be superfluous!*

What if a presidential candidate had the cojones to reply: "Would I favor the death penalty if my wife were raped and murdered? Equivocally not! I would find out what went wrong in the rapist-killer's family, that he would do such a terrible thing to my family. And then I would make it my goal in life to make sure that no one ever had to go through *that* with *their* family."

There it is. If you want to protect the family as a beacon of nurture, a bulwark of tender loving care, you don't do it by endorsing state-sponsored violence. You do it by taking values that safeguard the sanctity of family – bonding, belonging, empathy, nurture – and propagating them in ever-widening circles of inclusiveness until everyone has a home in the family of man.

Needless to say, that's the last thing the state wants. The business of family-building, of strengthening family values, of launching reconstruction efforts based on a familial way of being, of extending the bounds of nepotism by favoring everyone and his brother as a relative or friend, would inevitably come to fruition in the brotherhood of man. And the brotherhood of man would sound the death knell of the state as we know it and doom its entire power structure to obsolescence!

Implicit in the obscene question CNN's Bernard Shaw

POWERING DOWN CONFLICT AND WAR

posed to Dukakis is an old dark truth. The power of the state depends on the destruction of the family. Isn't that how it's always been? Isn't that why the culture hero, along with the institution of *kratos* or rule he embodies, becomes a force to be reckoned with through the destruction of the loving cocoon of the family?

Bereft of a sanctuary, how can we be free to feel? Devoid of the comfort and security of loved ones by our side, how can we stay alive to our feelings? That is the other way the state retains its death grip – by inducing emotional *rigor mortis*. By making us think we would sooner die than suffer the shame of opening up to our feelings. Drilled as we are in a Milgram curriculum, we know the score. Loveology majors are graded with a Scarlet Letter. Touchy-feely types who wear their heart on their sleeves wear a Yellow Star pinned to their chests. Shaming our manhood keeps us tough and aggressive enough to maintain an inner Department of Defense. We may make photo-op fools of ourselves by climbing into an Abrams battle tank, like Michael Dukakis, or striding on the deck of an aircraft carrier in a flight suit, like George W. Bush, but how else are we going to boost our credibility as *commander in chief, leader of the free world*?

Vulnerability to our pain, sensitivity to others' pain, are the crown jewels of the human race, but every generation pawns them to pay for a military-industrial complex. What need for a Star Wars defense when we can tap into feelings to keep from serving as an uninterrupted medium for the violent currents of the cosmos? Why draw Maginot lines in the sand against spooky gods and devils when we can mine our reserves of empathy to keep from serving as instruments of wrath for those who instruct with wrath? *Each individual possesses a conscience which to a greater or lesser degree serves to restrain the unimpeded flow of impulses destructive to others.*[1] Target that for extinction and we are all clear to relay the judgmental wrath of the gods and the punitive violence of the state.

It comes back to the same thing. In a trauma-crazed bid to combat the senseless violence of the universe, we become obedient vassals of gods. In a trauma-crazed bid to justify the senseless violence of gods, we disown our loving human sensibilities. Then we go ahead and fight the senseless violence of men by becoming submissive vessels of quasi-divine authorities who channel the senseless violence of gods! What a cosmic feedback loop we have gotten ourselves into!

To bust out we get our priorities straight. Before teaching the facts of life, we teach children the facts of humanity. *Superman is just someone orphaned at birth. If society were organized around dispensing our birthright of love, We the People would be the only superpower left standing.*

We lost our humanity once to the senseless violence of gods. Why lose it again to the senseless violence of men? We fight sins of our own making, crimes of our own device, by shutting down feeling centers and becoming unresisting conduits of a cosmic violence that was not even of our own doing. Pacifism isn't refusing to fight our enemies. It's refusing to fight our instinct to be friends. It's rejoicing in our touchy-feely senses, reinforcing our lovey-dovey nature, until we no longer channel the violence of gods or men. The only people-powered movement capable of world

---

[1] Stanley Milgram, **Obedience to Authority** [p. 188] Harper Torchbooks

revolution is the one that occupies our centers of emotional intelligence.

49. WE ARE NOT THE ENEMY! THE ENEMY IS FROM OUTER SPACE!

We live in a sea of separateness dotted with islands until we connect the dots. Jihadists are enemies of America for much the same reasons criminals are enemies of society. The Anglo-American conspiracy to despoil the collective resources of peoples can be broken down into the millions of daily acts that conspire to plunder our resources as individuals, usurp our reserves of living energy that could be put to good use fulfilling our life goals, and commandeer the time, talent, ingenuity and labor we need to fuel our dreams of a utopian planet. The orchestrated coup that puts a entire people under the despotic rule of a modern king of kings breaks down on the granular level into the billions of everyday acts that deny individuals the freedom to strike out on their own, exercise their initiative, follow their bliss and contribute to society in myriads of ways that uplift their dignity and worth.

When imperialists crush the hopes and dreams of entire hemispheres, people bristle with *anti-American* resentment. When parents, teachers, bosses collude in a domestic Operation Ajax to crush the hope and dreams of generations of youth, *they* bristle with *antisocial* resentment.

In the end, when this magma of rage and resentment erupts in volcanic blowback – when it explodes into fireballs that consume entire buildings and whole city blocks – it produces not just a single *night of terror* like Murray-Hill, but *an entire age of terror*!

In sum, the apocalyptic madness of the heavens is perpetuated not just through the madness of war but through the maddening legacy of power and

POWERING DOWN CONFLICT AND WAR

domination. The act of control is maddening in and of itself. Whether achieved by force or subterfuge, *it is like subjugating a woman through a rape that sows the wind and reaps the whirlwind.* For thousands of years the power structure has been stuck on an agenda of control that drives people mad. And then their maddened state becomes a pretext for imposing even more draconian levels of control – like locking them up in isolation cells where they are prone to hallucinations and wild mood swings. Control is maddening and madness in turn calls for even greater institutional forms of control! *The very thing we invoke to impose law and order is what drives people out of their minds – and their descent into insanity is a rationale for renewed demands for paramilitary force and martial law.*

To escape this age-spanning cycle of craziness we explode the myths of control that mess with our minds. We cure the mania for quasi-divine forms of control and people no longer have a reason for going crazy. Instead, they bring light and order to the world through the emotional intelligence of their nature.

The discovery of fire may have set mankind on the path to civilization, but the discovery of hellfire dragged it back to a dark ages. Kids not threatened with hell don't go around giving others hell, be it with furious words or sulfurous weapons. Kids raised without the rod don't have to lower their inner voice and raise a big stick. Remove the threat of guns as they do in Portugal and people treated with respect and dignity treat others the same way. Replace harsh degrading penal environments with luxury spas as they do in Norway and inmates don't have to riot like animals – they revert to human beings. Outlaw the rat race and skies once filled with signs of God's wrath are filled with cheers of jubilation at the Olympian lengths to which the human race goes to demonstrate its spirit of international cooperation and camaraderie.

So-called leaders are **purveyors of power.** In the process of repressing and suppressing our humanity, they cause us to lose our hearts and minds and souls. Parents are **purveyors of love.** In the seminal sense, they preside over uterine and extrauterine cocoons in which we gestate, grow, thrive, and come to fruition as fully developed human beings. In a remedial sense, they preside over havens of reparenting where we experience our buildup of negative feelings in the presence of empathic witness and advocates and release them to become whole once again.

The pivot of change is a no-brainer. We replace the **leadership model** with a **reparenting** one. We switch out those who lead us into submission and obedience for those who raise us up to flourish in a state of autonomy and self-realization. Leadership is mired in a mania for domination and control that is intent on monopolizing our hearts, souls, and minds. The worst thing is not all the bombing campaigns that vaporize the flesh. It is the rage for control and domination that disappears our humanity from the inside. Leaders *deself* us. They *desoul* us. They *discourage* us from functioning according to our lights. If we want a brave new world, we have to stop tinkering with a machinery of control that denatures and dehumanizes us, and pour our energy into a parenting model that procreates and empowers us to live our potential to the full. We replace the leader who is all about *disappearing* our humanity by dominating it through violence and force, with the **parent** who is all about bringing forth our humanity, drawing it forth with nurture and support, compassion and understanding, until it is **apparent** in all its shining glory!

In a speech before the United Nations in 1987, President Ronald Reagan spoke longingly of the world unity that would happen if aliens invaded Earth. "Perhaps we need some outside universal threat to make us recognize this common bond. I occasionally think how quickly our differences worldwide would vanish if we were facing an alien threat from outside this world."

We don't have to wish for an alien invasion of earth. *We just have to remember that one already happened.* Earth was invaded by gods of devouring fire from outer space. Our species was attacked by apocalyptic hosts that swooped from the heavens. Flaming pitch and brimstone poured from celestial ramparts as if the gods were bent on annihilating us. Not knowing any better, we incorporated these cosmic forces into human form and treated them like royalty. We embraced the crooked serpent in the sky and nursed it like a fiery viper in our bosom. *By allying with the wrath-crazed gods who have always been the real enemies of mankind, we made ourselves the enemy.* We became as a house divided into a warren of dynastic houses, each intent on dethroning the others in petty squabbles and internecine feuds, while the real menace has always lurked in the celestial background - the sinister white apparition of the comet, trailing death and destruction for humanity. We wasted aeons of time and worlds of energy fighting our own flesh and blood, when we could have unified ourselves to fight against being absorbed into the ranks of the walking dead by the arch nemesis that swarmed down from the frozen wastes of space!

Why be doormats for doomsday theologues or ideologues any more? To turn back the clock on an industrial revolution of death and destruction, we stop making the judgmental wrath of the gods the power plant of political order, stop legitimating it, vindicating it, weaponizing it. The thin blue line is not in line with the laws of human nature. *The harsh, punitive, authoritarian values that turn us against each other come not from the Bible Belt but the Kuiper belt.* The deadliest of all addictions has always been our addiction to the wrath of God. We quit carrying the fire of extraterrestrial invasions and burning each other with it. As the mass death of our species cannot be justified by any moral calculus, we cease striving for our mutual extinction. As the music never stops once we play in tune with our heartstrings, we leave off playing glorified games of musical chairs like the Great Game or Game of Thrones. *We come together, hold each other, reverently gaze into our souls until the teardrops fall from billions of eyes and coalesce into a swelling tsunami of grief that extinguishes the fires of celestial wrath once and all.*

Now that we know what to fight, we can be on the right side of history by championing values instinct in our psychobiology. We can fight the brute forces of cosmic nature by tenderly nursing our psyches back to a consciousness of our humane nature. We can defeat the *insentient* hostilities that came from outer space by *sensitizing* ourselves to our inner space. We can take the pent-up anguish and rage we faithfully recycled down the generations and release it in a universal outpouring of love. And then the only thing better than that time long ago when we got our beautiful planet back would be getting back our magnificent humanity right here and now!

Made in the USA
San Bernardino, CA
28 March 2019